SpringerBriefs in Electrical and Computer Engineering

Speech Technology

Series Editor

Amy Neustein

For further volumes:
http://www.springer.com/series/10043

Editor's Note

The authors of this series have been hand selected. They comprise some of the most outstanding scientists—drawn from academia and private industry—whose research is marked by its novelty, applicability, and practicality in providing broad-based speech solutions. The Springer Briefs in Speech Technology series provides the latest findings in speech technology gleaned from comprehensive literature reviews and empirical investigations that are performed in both laboratory and real life settings. Some of the topics covered in this series include the presentation of real life commercial deployment of spoken dialog systems, contemporary methods of speech parameterization, developments in information security for automated speech, forensic speaker recognition, use of sophisticated speech analytics in call centers, and an exploration of new methods of soft computing for improving human–computer interaction. Those in academia, the private sector, the self service industry, law enforcement, and government intelligence are among the principal audience for this series, which is designed to serve as an important and essential reference guide for speech developers, system designers, speech engineers, linguists, and others. In particular, a major audience of readers will consist of researchers and technical experts in the automated call center industry where speech processing is a key component to the functioning of customer care contact centers.

Amy Neustein, Ph.D., serves as editor in chief of the International Journal of Speech Technology (Springer). She edited the recently published book Advances in Speech Recognition: Mobile Environments, Call Centers and Clinics (Springer 2010), and serves as quest columnist on speech processing for Womensenews. Dr. Neustein is the founder and CEO of Linguistic Technology Systems, a NJ-based think tank for intelligent design of advanced natural language-based emotion detection software to improve human response in monitoring recorded conversations of terror suspects and helpline calls.

Dr. Neustein's work appears in the peer review literature and in industry and mass media publications. Her academic books, which cover a range of political, social, and legal topics, have been cited in the Chronicles of Higher Education and have won her a pro Humanitate Literary Award. She serves on the visiting faculty of the National Judicial College and as a plenary speaker at conferences in artificial intelligence and computing. Dr. Neustein is a member of MIR (machine intelligence research) Labs, which does advanced work in computer technology to assist underdeveloped countries in improving their ability to cope with famine, disease/illness, and political and social affliction. She is a founding member of the New York City Speech Processing Consortium, a newly formed group of NY-based companies, publishing houses, and researchers dedicated to advancing speech technology research and development.

Ami Moyal • Vered Aharonson
Ella Tetariy • Michal Gishri

Phonetic Search Methods for Large Speech Databases

 Springer

Ami Moyal
Afeka Academic College
   of Engineering
Tel-Aviv, Israel

Vered Aharonson
Afeka Academic College
   of Engineering
Tel-Aviv, Israel

Ella Tetariy
Afeka Academic College
   of Engineering
Tel-Aviv, Israel

Michal Gishri
Afeka Academic College
   of Engineering
Tel-Aviv, Israel

ISSN 2191-8112   ISSN 2191-8120 (electronic)
ISBN 978-1-4614-6488-4   ISBN 978-1-4614-6489-1 (eBook)
DOI 10.1007/978-1-4614-6489-1
Springer New York Heidelberg Dordrecht London

Library of Congress Control Number: 2012956552

Printed on acid-free paper

Springer is part of Springer Science+Business Media (www.springer.com)

Preface

A leading use of speech recognition technology is the conversion of large speech databases into text for indexing and retrieval purposes. Using a large vocabulary continuous speech recognition (LVCSR) engine seems to provide a natural solution, as speech can be fully converted into text and then indexed and searched.

One method used for searching speech databases is Keyword Spotting (KWS). Speech recognition technology is used in KWS-based applications to enable specific words to be identified out of a stream of continuous speech. This is particularly useful when a relatively small number of words need to be quickly pinpointed within a huge speech database.

KWS can be implemented using various methods. The phonetic search approach is presented together with an analysis of its computational complexity. Following which, an anchor-based phonetic search algorithm is presented with evaluation results of its computational complexity. The KWS recognition performance using the anchor-based search is compared to an exhaustive search on several speech databases.

The purpose of this brief is to present the challenges involved in performing phonetic search KWS in large speech databases, with a specific focus on efficient searching. Ideally, all the underlying algorithms and related topics would have been presented, however this would be incongruent with the value of a "brief." Thus in compensation, various published works were referred to in cases where additional information may be helpful to the reader.

Our research is currently focused on phonetic search based KWS within a lattice of phonemes and an extension of the search to multiple lattices generated from several languages in order to support KWS in languages with limited language resources.

Tel-Aviv, Israel

Ami Moyal
Vered Aharonson
Ella Tetariy
Michal Gishri

Acknowledgments

The underlying anchor-based algorithm research reported here was partially supported by grants funded by the chief scientist of the Israeli Ministry of Commerce. The original research targeted the reduction of the active vocabulary used by a speech recognition engine (Tetariy et al. 2010). We took this approach and applied it to efficient keyword spotting in large speech databases.

We thank the Afeka Academic College of Engineering for institutional support and especially to Prof. Moti Sokolov, the college president, who has believed in and supported our activities from day 1.

Our appreciation to Springer for providing us with the opportunity to publish this brief.

Contents

Chapter 1
Keyword Spotting Out of Continuous Speech

1.1 Introduction

Successful Automatic Speech Recognition (ASR) technology has been a research aspiration for the past five decades. Ideally, computers would be able to transform any type of human speech into an accurate textual transcription. Today's ASR technology generates fairly good results using structured speech with relatively low Signal to Noise Ratios (SNR), but performance degrades when using spontaneous speech in real-life noisy environments (Murveit et al. 1992; Young 1996; Furui 2003; Deng and Huang 2004). Performance that is acceptable for commercial applications can be achieved using large training corpora of speech and text. However, there are still problems that need to be resolved.

One of the main problems is the mismatch between training and testing (real-life) conditions (Young 1996; Baker et al. 2009; Tsao et al. 2009; Furui et al. 2012; Saon and Chien 2012). Types of mismatches include: background noise, channel distortion, Out of Vocabulary (OOV) words (when speakers use words not in the recognition vocabulary), foreign accent speech, etc. Various methods and algorithms for minimizing this mismatch between training and testing have been suggested and implemented (Mammone et al. 1996; Sankar and Lee 1996; Huo et al. 1997; Matrouf and Gauvain 1997; Viikki and Laurila 1998; Hirsch and Pearce 2000; Barras et al. 2002; Parada et al. 2010; Kai et al. 2012), while in parallel, larger amounts of representative speech (usually from live deployments) have been injected into the training process using automatic procedures that do not necessitate manual transcription of the data (Kamm and Meyer 2002; Evermann et al. 2005; Heigold et al. 2012).

The leading approach in ASR today is searching for the most probable sequence of words that describes the input speech. The search uses: (1) acoustical models representing the phonemes of the target language; (2) a lexicon of the recognition vocabulary words represented as sequences of phonemes; and (3) a Language Model (LM) specifying the word transition probabilities. ASR is performed by inputting a sequence of vectors estimated from the input speech signal to the

A. Moyal et al., *Phonetic Search Methods for Large Speech Databases*, SpringerBriefs in Speech Technology, DOI 10.1007/978-1-4614-6489-1_1, © The Author(s) 2013

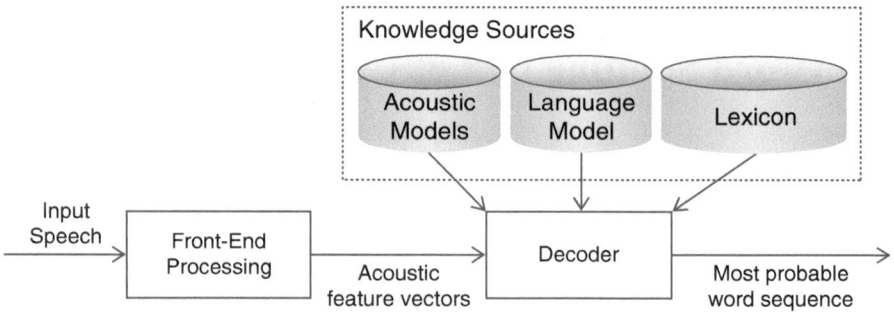

Fig. 1 Speech recognition engine

engine, and then using the combined information from the knowledge sources – the acoustical models, lexicon and LM – to search for the most probable sequence of words. A high level description of a speech recognition engine is illustrated in Fig. 1.

The search for the most probable sequence of words can be represented using the following notation:

$O = \{o_1, \ldots, o_T\}$ – A sequence of vectors representing the speech signal (the output of the front-end processing stage).

$W = \{w_1, \ldots, w_N\}$ – A hypothesized sequence of words.

$P = \{P_1, \ldots, P_M\}$ – A sequence of phonemes representing the hypothesized sequence of words.

$S = \{S_1, \ldots, S_R\}$ – A sequence of phoneme states representing a sequence of phonemes.

The search algorithm looks for the most probable sequence of words for a given sequence of vectors over all possible W:

$$MAX\{p(W|O)\} \tag{1.1}$$

The probability of a hypothesis that observes a sequence vector O and recognizes a word sequence W can be represented using Bayes' rule as:

$$p(W|O) = \frac{p(O|W)p(W)}{p(O)} \tag{1.2}$$

The probability p(O) can be removed from the search for maximum probability since it is a common term to all hypotheses. The probability p(W) is the a priori probability of receiving a word sequence W that can be computed using N-gram probabilities.

The goal is to estimate the following probability:

$$p(O|W) = \sum_{S,P} p(O,S,P|W) \tag{1.3}$$

where the sum is over all possible phoneme sequences P, and all possible state sequences S.

However, the number of possible paths grows exponentially for both S and P. Thus a suboptimal solution leads to the following commonly used approximation:

$$p(O|W) \approx \text{MAX}_{S,P}\, p(O|S,P,W)p(S|P,W)p(P|W) \tag{1.4}$$

Under the factorization property of the given joint probability:

$$p(O|S,P,W) = p(O|S) \tag{1.5}$$

and

$$p(S|P,W) = p(S|P) \tag{1.6}$$

The following equation can be reached:

$$p(O|W) \approx MAX_{S,P}\, p(O|S)p(S|P)p(P|W) \tag{1.7}$$

where

$p(O|S)$ is the probability of receiving a vector sequence O for a given state sequence S.
$p(S|P)$ is the probability of receiving a state sequence S for a given phoneme sequence P.
$p(P|W)$ is the probability of receiving a phoneme sequence P for a given word sequence W.

p(O|S) is estimated using the acoustical models (usually a mixture of Gaussians representing an HMM state). $p(S|P)$ can be estimated using HMM state transition probabilities. $p(P|W)$ is a sequence of phonemes resulting from transformation of the word sequence W into a phoneme sequence using a pronunciation lexicon.

Thus the search for the most probable sequence of words is basically a search for the most probable sequence of phonemes, or more accurately, a search for the most probable sequence of phoneme states by maximizing the probability p(O|W) (Gosztolya and Tóth 2011).

The performance required from the above mentioned algorithms highly depends on the intended application. Currently, the two most common uses of ASR technology are in Human-Machine Interaction (HMI) and speech database (DB) indexing and retrieval.

HMI applications include call center self-service, personal assistance, directory assistance, dictation, call routing, etc. In these types of applications, the active recognition vocabulary can range from dozens of words (small vocabulary) to tens of thousands of words (very large vocabulary). The input speech may consist of short utterances of only a few words, such as: "Pay my bill" or of much longer utterances such as: "I would like to order a flight from New York to London for next Sunday." The recognition can be controlled by a structured grammar that defines all the potential utterances or by a language model presenting statistical probabilities of word transitions. Regardless of which alternatives are employed, these types of applications require real-time full transcription of the input speech.

Speech indexing and retrieval is crucial in today's world of massive amounts of digital data, including speech and multi-media that require searching capabilities. ASR technology is often used to convert large speech databases into text as a pre-processing stage (Witbrock and Hauptmann 1997; Clements et al. 2002; Thambiratnam 2005; Burget et al. 2006; Mamou et al. 2008; Szöke et al. 2008; Baker et al. 2009; Schneider 2011; Saon and Chien 2012). In this case, a Large Vocabulary Continuous Speech Recognition (LVCSR) engine seems to provide a natural solution, as speech can be fully converted into text and then indexed so that a search request can be responded to. The quality of the search results, however, depends heavily on the accuracy of the speech recognition engine. Furthermore, while the textual transformation and indexing can be performed off-line, response time to the search request in most applications is crucial. While users often expect fast and even real-time responses, the actual time it takes to respond to a search query depends on the computational complexity of the search, which in turn is dependent on the size of the speech database.

One method used for searching speech databases is Keyword Spotting (KWS). Speech recognition technology is used in KWS-based applications to enable specific words to be identified out of a stream of continuous speech (Wilpon et al. 1990). This is particularly useful when it is necessary to quickly pinpoint a relatively small number of words within huge amounts of speech data. The technology is often used by call centers and security-intelligence organizations for categorizing calls and searching speech databases, and by companies offering multi-media search applications in the internet and enterprise markets.

KWS can be performed using one of three possible methods (Szöke et al. 2005):

1. **LVCSR based KWS** – A Large Vocabulary Continuous Speech Recognition (LVCSR) engine produces a transcription of the entire speech database and the KWS-based application searches the resulting text for the designated keywords (Cardillo et al. 2002; Szöke et al. 2005, 2008; Burget et al. 2006; Šmídl and Psutka 2006; Dubois and Charlet 2008; Wang et al. 2008; Motlicek et al. 2012).
2. **Acoustic KWS** – The KWS engine operates on the speech itself and the recognition vocabulary consists only of the designated keywords (Szöke et al. 2005; Motlicek et al. 2012).
3. **Phonetic Search KWS** – A phoneme recognition engine produces a phonetic transcription of the entire speech database and a phonetic search engine searches the resulting textual phoneme sequence (or lattice) for the designated keywords (Amir et al. 2001; Szöke et al. 2005; Burget et al. 2006; Thambiratnam and Sridharan 2007; Bar-Yosef et al. 2012).

Each method has its advantages and shortcomings (see Sect. 2.4), making one more suitable than the others, depending on the specific application. However, because KWS often runs on extremely large speech databases, the issue of computational complexity is vital regardless of the target application.

1.2 Problem Formulation: KWS in Large Speech Databases

From a high level perspective, KWS in large speech databases can be considered a classical search problem. Specific words or terms (keywords) are searched for within a given database that generally consists of many speech files (recorded calls, lectures, video sound tracks, etc.). The expected search result is an ordered list of files with precise time pointers to the locations where any one of the keywords has been spotted. When it comes to KWS, a full transcription of the entire speech database is not always required, as it might be for other applications of speech recognition. The main issue is the rapid identification of the keywords in their accurate locations. These keywords can then serve as pointers to specific data that will most probably be further analyzed (manually or automatically).

Crucially, these search results may contain not only correctly recognized words, but also what is known as false alarms. False alarms are keyword recognitions in locations where a keyword does not actually exist in the speech. The tradeoff between correct recognitions and false alarms is an important issue. The distribution of both is measured as a function of a distance threshold in order to determine the optimal working point. The distance measure is computed for each keyword hypothesis, but the keyword is declared recognized only if its computed value is lower than the a priori threshold. The optimal working point is selected based on the requirements of the specific KWS application, and is crucial to its success. An application that produces a large number of false alarms will degrade the quality of the search results and lead to tedious filtering of false alarms by the user.

Usually, the detection and false alarm rates are the main KWS performance characteristics that are directly reflected in the search results presented to the user. However, it is worth mentioning that missed detections are not part of the output of a KWS system. These missed detections are actual occurrences of keywords within the speech database that were not spotted by the system. However, when evaluating a KWS system, the missed detection rate can be calculated, as it is the direct complement of the detection rate.

Another important specification of a KWS-based application is the response time, particularly when searching in large speech databases. The response time requirement is highly dependent on the market segment or application and varies from real-time to a few minutes or even hours. Much effort has been invested in reducing the search complexity of the KWS algorithm itself, but response time is also dependent on the classical indexing and retrieval methods used on the text.

1.3 Target Applications of Keyword Spotting

Basically, any application that requires extracting specific data from large speech or multimedia databases is suitable for employing KWS. In most cases, the main motivation is to scan a large speech or multimedia database and locate all files that contain a specific word for further analysis (manually or automatically).

For example, speech analytics platforms offered to the call center market may incorporate KWS for tracking down calls between agents and customers. The list of words to be searched is tailored to the needs of the customer (call center) and may include: product names for analysis of customer responses to new products, customer opinion terminology for tracking customer satisfaction, or competitor names for facilitating customer retention, etc. (Alon 2005; Mishne et al. 2005; Park et al. 2008). Only calls containing the designated keywords are forwarded to the appropriate department for further analysis. Detailed analysis is both human-resource and time consuming, as it may include listening to recordings, automatic topic classification or even full transcription by an LVCSR engine. Thus KWS is a crucial pre-analysis stage that filters out irrelevant data and allows companies to focus their efforts.

Similarly, KWS applications are used by the security intelligence market for surveillance and monitoring applications (Alon 2005; Thambiratnam 2005; Wallace et al. 2007). KWS can be used to filter out many hours of irrelevant data and transfer only recordings tagged with vital security information for further analysis. Also important for the security market is the ability to change the keyword list on-the-fly due to ever-changing security warnings, which often include foreign names and words (Thambiratnam and Sridharan 2005).

Another potential market for KWS applications is the internet search market where KWS can be used to automatically index multi-media content, such as podcasts, video soundtracks, etc. Extracted keywords can serve as textual metadata to be indexed by search engines (Cardillo et al. 2002), thus enabling search results to include multimedia content in addition to textual content. Search results can also include pointers to the exact location of the spoken word within the multimedia content.

Unlike the use for intelligence purposes (security or business), where in most cases off-line processing is required, internet searches generally require real-time response rates and usually work on much larger databases. KWS in internet searches can also be applied by the enterprise and media markets, which may not require real-time responses and are performed on smaller speech databases.

Chapter 2
Keyword Spotting Methods

This chapter will review in detail the three KWS methods, LVCSR KWS, Acoustic KWS and Phonetic Search KWS, followed by a discussion and comparison of the methods.

2.1 LVCSR-Based KWS

Performing KWS on textual databases is relatively straightforward. The text is perused for a given list of words and the location of the words is tagged within the text. Translating this method for use in speech databases is a two-stage process. First, an LVCSR engine is employed to transform the entire speech signal into text. The LVCSR engine performs the search for the most probable sequence of words based on the Viterbi search algorithm, using acoustic models, a large lexicon of words and a language model. In the second stage, the KWS mechanism utilizes established text-based search methods to locate the keywords within the text. An indexing phase can be performed on the resulting text in order to accelerate the search response time. This method will be referred to as LVCSR-based KWS.

Figure 2 illustrates the two sequential stages involved in LVCSR-based KWS.

2.2 Acoustic KWS

Another common KWS method is Acoustic KWS. Using this method, the engine does not attempt to transcribe the entire stream of speech. Like the LVCSR-based method, this method employs the Viterbi search. That is, the system employs a speech recognition engine on the speech. However, rather than a large vocabulary which is intended to cover all potentially spoken words, a smaller set of designated keywords is used as the recognition vocabulary (Thambiratnam 2005) and general speech models (as part of the acoustical models) are used to model

A. Moyal et al., *Phonetic Search Methods for Large Speech Databases*, SpringerBriefs in Speech Technology, DOI 10.1007/978-1-4614-6489-1_2, © The Author(s) 2013

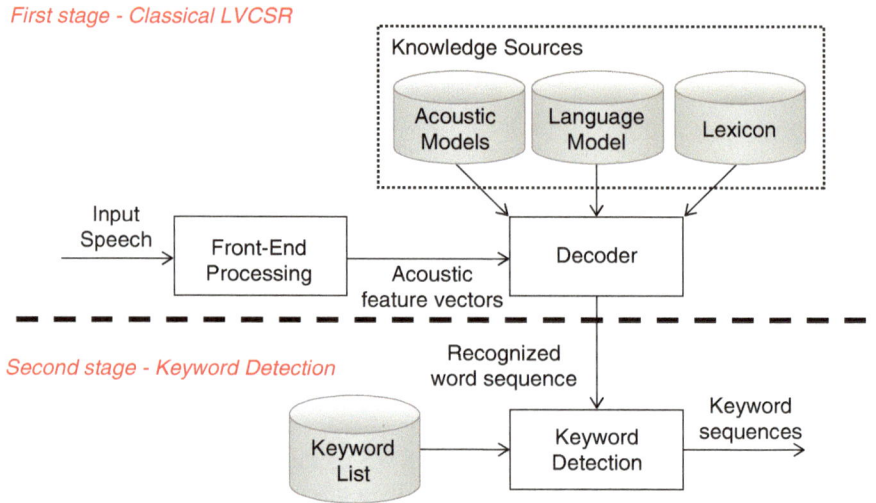

Fig. 2 An LVSCR keyword spotting system – one-time transformation of a speech database (DB) into a textual word DB and KWS engine

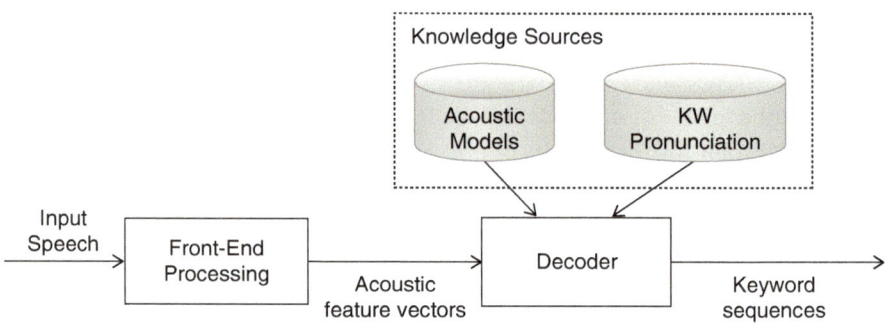

Fig. 3 An acoustic keyword spotting system

non-keyword speech (Szöke et al. 2005). Thus, acoustic KWS can be performed in only one stage, as illustrated in Fig. 3.

2.3 Phonetic Search KWS

As its name suggests, phonetic search KWS utilizes a phonetic search engine. In the first stage, a phoneme decoder is employed once to transform the speech input into a textual sequence. However, rather than producing a string of words, the decoder transforms the speech signal into a string (or lattice) of phonemes (Amir et al. 2001;

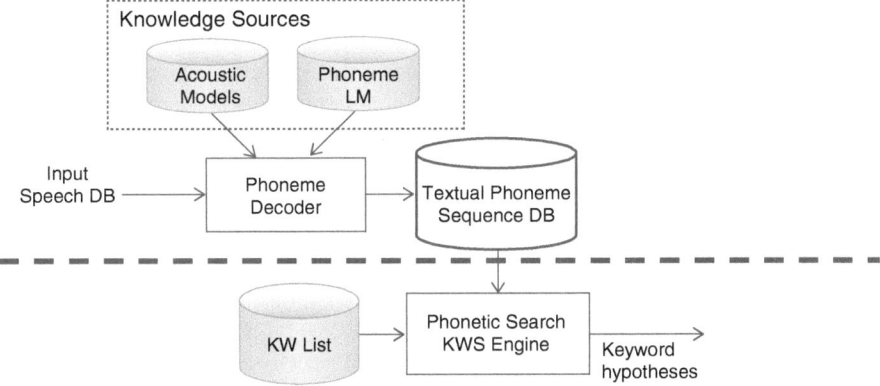

Fig. 4 A phonetic search system – one-time transformation of a speech DB to a textual phoneme DB and KWS phonetic search engine

Yu and Seide 2004; Thambiratnam and Sridharan 2005). In the second stage, the phonetic search engine employs a distance measure to compute the textual distance between the phoneme sequences that correspond to the keyword vocabulary and the phoneme sequences within the phoneme string (Alon 2005).

As shown in Fig. 4, the phonetic search engine uses two types of input data: a list of keywords, where each word is represented by a sequence of phonemes, and a speech database which has been run through a phoneme decoder to produce a sequence of recognized phonemes.

2.4 Discussion: Why Phonetic Search?

Each of the three KWS methods presented above has advantages and shortcomings. The crucial parameters to evaluate are response time, KWS performance, and keyword flexibility (James and Young 1994; Dharanipragada and Roukos 2002; Mamou et al. 2007; Thambiratnam and Sridharan 2007; Schneider 2011).

2.4.1 Response Time

In terms of overall computational complexity, LVCSR-based KWS and phonetic search KWS both implement a double stage process: (1) transformation of speech to text (word sequences in the case of LVCSR and phoneme sequences in the case of phonetic search) and (2) a keyword search (word-based in the case of LVCSR and phoneme-based in the case of phonetic search). Acoustic-based KWS, on the other hand, is performed in one stage and operates on the speech itself with no textual transformation. Although a keyword search that is implemented on fully transcribed text in the LVCSR method is fast (particularly if the text has also been indexed), it is usually at a disadvantage in comparison to the phonetic search and acoustic methods due to the fact that an LVCSR engine demands a large vocabulary and a complex language model to produce recognition results, thus resulting in a high level of complexity during the pre-processing stage.

The phonetic search method performs phoneme recognition using phoneme transition probabilities (di-phones) with no lexicon or word level language model. During the search stage, however, phonetic search KWS uses a textual sequence distance measure that requires more computation. This is because the phonetic search must generate word-level hypotheses based on phoneme sequences, while in LVCSR-based KWS the textual output is already word-level (Burget et al. 2006).

In contrast, the acoustic-based KWS uses a vocabulary consisting only of the keywords and does not require a language model at all. Because the acoustic-based method operates on the speech itself and requires only a small vocabulary, it is appropriate for real-time keyword spotting or KWS in small speech databases. However, this means that general speech must be well-modeled (Thambiratnam 2005) to avoid extensive over detection (false alarms).

2.4.2   KWS Performance

The spontaneous speech and poor recording quality of speech databases often leads to deficient LVCSR performance (Butzberger et al. 1992; Cardillo et al. 2002). The large number of disfluencies, including mispronounced words, false starts, filled pauses, overlapping speech, speaker noises and background noise found in spontaneous speech (Butzberger et al. 1992; Gishri and Silber-Varod 2010) often results in outputs strewn with word insertions, deletions and substitutions. Thus the "most probable" word sequences produced by the engine may not adequately reflect the actual input speech. This, in turn, affects the reliability of the keyword search.

The same is true with regard to phonetic search results. Poor phoneme recognition may yield lower keyword recognition performance in comparison with the acoustic KWS method, which works on the speech itself by searching for a specific sequence of phonemes without textual transformation.

2.4.3   Keyword Flexibility

In comparison to the phonetic search method, which runs on sequences of phonemes rather than words, the LVCSR method is at a disadvantage when it comes to keyword flexibility (Cardillo et al. 2002; Burget et al. 2006; Wallace et al. 2007). Using the phonetic search method allows application users total freedom in changing the designated keywords, since the textual transformation into phonemes is not restricted by a vocabulary. Adding new keywords is a simple procedure that entails re-running the phonetic search on the phoneme sequences, but does not require re-running the phoneme decoder.

The textual transformation produced by an LVCSR engine, on the other hand, is constrained by the recognition vocabulary and the language model employed. Thus, unless the designated keywords were part of the original recognition vocabulary, they cannot be changed without repeating the recognition process (Clements et al. 2001; Cardillo et al. 2002; Szöke et al. 2005; Mamou and Ramabhadran 2008). Since keywords are in many cases names or domain-specific vernacular, they are often not found in standard lexicons (Wallace et al. 2007; Gishri and Silber-Varod 2010). This is a substantial shortcoming of the LVCSR method.

Acoustic-based KWS also represents an impractical solution for searching large databases that require rapid and flexible searching capabilities. Because it consists of only one stage, the entire process needs to be re-run on the speech database each time a new keyword dictionary is introduced.

The majority of applications require keyword flexibility, as well as the shortest possible response time when searching very large speech databases, making the phonetic search KWS method more attractive than the LVCSR and acoustic-based options when searching very large speech databases. Thus, the focus of the following chapters will be on phonetic search KWS, and the implementation of an algorithm for the reduction of computational complexity in the phonetic search KWS process.

Chapter 3
Phonetic Search

Traditionally, information retrieval techniques create an index of words or terms found in a textual database that can later be rapidly searched by simply entering a query for a desired word or term. Naturally, straightforward application of this technique to non-textual materials is impossible. When it comes to speech, using text-based techniques requires a preprocessing stage of transforming the digital speech signal into some form of text. However, since classical speech recognition engines are not totally accurate, the indexing will necessarily include errors.

Phonetic search mechanisms aim to reduce the inaccuracy of the textual transformation by taking one step back. Rather than transforming speech into a full transcription of spoken words, the transformation stops at the phoneme level, thus no rigid decisions about word bindings need to be made (Cardillo et al. 2002).

The result is a phonemic representation of the entire speech database; thus, search queries should also be phonemic representations of the desired word or term. These are typically generated by an automatic transformation (grapheme-to-phoneme and/or lexicon based) of the user keyword input. The advantage is that the match between the search term and phoneme string found in the database does not have to be exact, nor does the orthographic spelling of the word (Cardillo et al. 2002). The phonetic search mechanism can provide a list of near matches from the database, thus potentially overcoming errors in the textual transformation.

3.1 The Search Mechanism

As described above, phonetic search KWS is implemented in two sequential stages: the first consists of a transformation of the speech database into phoneme sequences and the second is the actual keyword search.

The first stage is performed using a phoneme decoder which converts the speech database into phonemes. The dataset used for the keyword search in the second stage is thus a phonemic representation of the original digital speech, rather than words. This initial transformation of speech into phonemes is a one-time, fast, and

A. Moyal et al., *Phonetic Search Methods for Large Speech Databases*, SpringerBriefs in Speech Technology, DOI 10.1007/978-1-4614-6489-1_3, © The Author(s) 2013

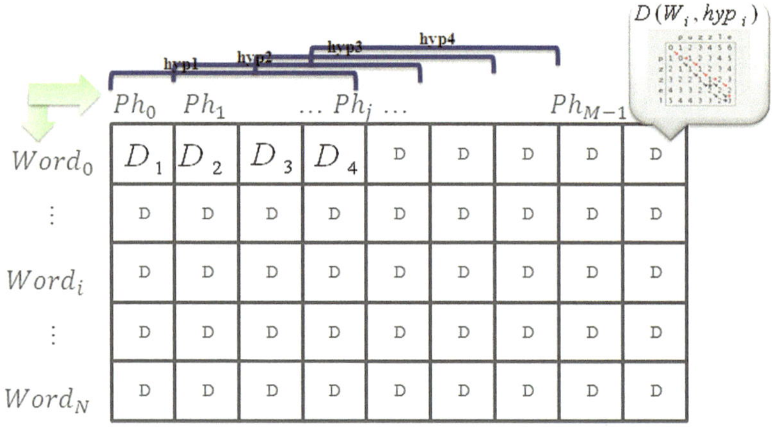

Fig. 5 A phonetic search space

usually off-line, procedure (Wallace et al. 2007). It will thus not be addressed further. Rather, the focus will be on the second stage – the phonetic search. An emphasis will be placed on the computational complexity of the search itself, which is a major concern as it dictates the response time to a search request.

The keyword search is performed by exact or near-matching phonemic sequences representing the keywords (produced either manually or automatically) to phonetic sequences found in the database. The matching process is performed using a measure that calculates the distance between the keyword phoneme sequence and a hypothesized sequence from the database.

The search space can be defined as a grid, G (see Fig. 5). Given an input list of N keywords and a sequence of recognized textual phonemes of length M, the size of G is NxM. Each cell in the grid contains the distance between W_i (a keyword represented by its phonemic transcription) and a hypothesis (a partial textual phoneme sequence from the database that begins with the recognized phoneme Ph_j and has the same phoneme length as W_i).

An exhaustive search entails the construction of all possible hypotheses for matching. That is, each keyword can be considered as possibly occurring in any location of the sequence of phonemes in the database.

Thus, each keyword W_i of length L that starts with the phoneme Ph_j produces $M - L + 1$ different hypotheses:

$$Hyp[j,\ L] = \{Ph_j,\ Ph_{j+1}, \ldots, Ph_{j+L-1}\} \quad \text{where} \quad 0 \leq j \leq M - L + 1 \qquad (3.1)$$

Each keyword is assigned a length in terms of the number of phonemes:

$$L_i = Length\{W_i\} \qquad (3.2)$$

Then the value of each cell in the search space is computed by:

$$G[i, \; j] = Distance\{W_i, \; Hyp[j, \; L_i]\} \qquad \text{where } 0 \leq i \leq N, \qquad (3.3)$$

$$0 \leq j \leq M - L_i + 1$$

The distance is usually computed using the "Edit Distance," also known as "Levenshtein Distance" (Pucher et al. 2007; Hermelin et al. 2009). This distance measure is used to compare the phoneme string associated with a given keyword to the phoneme string resulting from the speech database. The distance is actually the minimal sum of differences between the two sequences. Possible differences between the two phoneme sequences are defined as phoneme substitutions (the keyword contains one phoneme while the recognized string contains another), deletions (the keyword contains a phoneme while the recognized string is missing a phoneme) and insertions (the keyword does not contain a phoneme while the recognized string does). A specific weight can be associated with each of the transformations when computing the distance measure. Usually insertions, deletions and substitutions are assigned an equal weight of 1/3, while correct phonemes are assigned a weight of zero.

The output of this matching process is generally a list of occurrences where a match is defined as "close-enough" by an a priori threshold. Each keyword alert is presented with its distance measure value and location in the speech database (referred to as a keyword hypothesis).

3.2   Using Phonetic Search for KWS

The advantages of phonetic search have been demonstrated in several studies and implementations of KWS systems. Burget et al. found LVSCR and phonetic search KWS results to be comparable, but also highlighted the necessity to accelerate the search phase when implementing a phonetic search (Burget et al. 2006). In searching large Digital Media Asset Management (DMAM) systems, Cardillo et al. found phonetic searching to be faster than the LVCSR method, and much more accurate when using longer search terms (Cardillo et al. 2002). Moreover, Bar-Yosef et al. demonstrated results searching databases in languages with only limited training resources (Bar-Yosef et al. 2012). In such cases, phonetic decoding may be the only option in the absence of a thoroughly trained LVCSR engine (Wallace et al. 2007).

However, phonetic search also has limitations. It is prone to high levels of false alarms, especially with short keywords (Cardillo et al. 2002; Wallace et al. 2007). Since queries are ranked according to their distance measure values, false positive detections are imminent; a segment that does not match the keyword may be retrieved as a result of a lower distance measure value than the set threshold,

while a segment that does match may be rejected due to a distance measure value
higher than the set threshold.

This problem can be moderated by modifying the threshold for the distance
measure. However, threshold setting carries an eternal trade-off dilemma; a high
threshold may decrease the number of false alarms produced, but at the cost of
missing true occurrences of the keyword (an increase in the missed detection rate).
A common operating guide is to allow a pre-defined number of false alarms per
hour of speech. The overall rate of false alarms then depends on the fidelity of the
algorithm.

When it comes to practical implementations of keyword spotting, application
users generally have two main requirements; that the keyword spotting is
vocabulary-independent and that it is rapid.

Phonetic search KWS meets the first requirement (see Sect. 2.4). However,
standard mechanisms are computationally cumbersome (Thambiratnam and
Sridharan 2007; Wallace et al. 2007; Har-Lev et al. 2010), particularly when
using a lattice output at the phoneme decoder stage in order to increase the number
of keyword spotting hypotheses. In phonetic search KWS, speech data is pre-
processed only once, while the search phase may be repeated indefinitely for new
keywords. Thus, the efficiency of the search is crucial (Vergyri et al. 2007).

3.3 Computational Complexity Analysis

As described in Sect. 3.1, the distance calculation used for a phonetic search allows
for deletion, insertion, and substitution of phonemes. The resulting complexity is
$O(Average(L)^2)$, where L refers to length (in phonemes) of a keyword. The
Levenshtein distance grid for generating each hypothesis is shown in Fig. 6.

The value in each cell of the Levenshtein distance grid is computed as the
minimal number of phoneme substitutions, insertions and deletions needed to
transform one phoneme string into the other, where each substitution, insertion or
deletion generates a "cost" to the accumulated distance measure, while correct
recognition of a phoneme generates no cost. These costs can be calculated during a
preprocessing analysis of the confusion matrix (Gusfield 1997) or assigned a 1/3
value for the three types of errors and 0 for correct. The distance is calculated using
a dynamic programming process as follows:

$$
M[i,j] = Min \begin{cases} M[i-1,j] + c_{Del}(Ph_i) \\ M[i,j-1] + c_{Ins}(Ph_j) \\ M[i-1,j-1] + c_{Sub}(Ph_i,Ph_j) \\ M[i-1,j-1], \ (if \ Ph_i = Ph_j) \end{cases} \quad (3.4)
$$

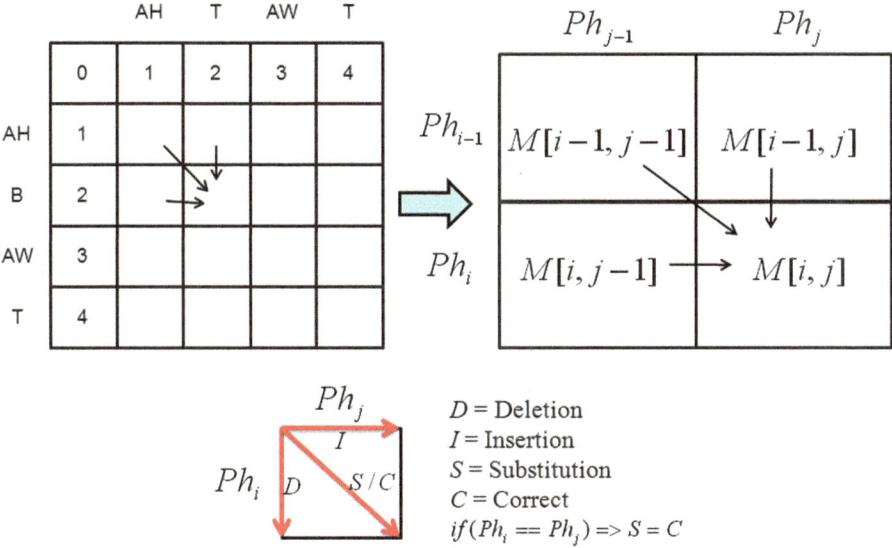

Fig. 6 Basic structure of a Levenshtein grid

where $c_{Del}(i)$ signifies the cost if Ph_i was deleted in the output of the engine; $c_{Ins}(j)$ signifies the cost if Ph_j was inserted in the output of the engine; and $c_{Sub}(i,j)$ signifies the cost if input Ph_i was replaced by Ph_j in the output of the engine.

In summary, the overall computational complexity of this exhaustive phonetic search is $O(NM(Average(L))^2)$.

Since phonetic search KWS generates a search space that is linear to the number of keywords and the length in phonemes of the speech utterances in the searched database, the resulting level of computational complexity is unacceptable for KWS-based applications that perform searches on very large speech databases and that require real-time responses. It is therefore necessary to reduce the computational complexity involved in the phonetic search KWS process, without adversely affecting performance.

Chapter 4
Search Space Complexity Reduction

4.1 Overview

Various studies have focused on exploring ways to search more efficiently; this chapter will present an overview of methods that deal with efficient searching, with a focus on methods that reduce the size of the search space. The basis of all these methods is to formulate and use constraints that trim down the search space by eliminating impossible paths, dimensions or locations, thus leaving a reduced grid on which to perform the search.

The issues of dimensionality reduction and similarity search have been extensively addressed in textual information retrieval and textual data mining literature (Bingham and Mannila 2001; Kuri-Morales and Rodríguez-Erazo 2009). The reduction mechanism varies between studies, and ranges from arbitrary to knowledge-based. Searching in these domains is performed on written letters, words and sentences governed by grammar. Conceptually, this is not far from textual phoneme sequences and thus many insights can be derived from these studies.

In speech related technologies, various search methods have been investigated with the main motivation of reducing the overall search space. The two main approaches involve either splitting the search operation into two stages (where possible) or reducing the number of possible search paths.

A leading method in constraining the search space is employing an "initial pass" scheme where the search space is constrained by a coarse match to part of the knowledge sources. The search itself can be then performed on this reduced search space using all the available knowledge sources. Of course, an overall perspective should take into account the total complexity of the process (initial pass and the search itself) but other aspects need to be considered as well. For example, the initial pass can be performed off-line as a preliminary stage that prepares a reduced and structured search space.

A. Moyal et al., *Phonetic Search Methods for Large Speech Databases*, SpringerBriefs in Speech Technology, DOI 10.1007/978-1-4614-6489-1_4, © The Author(s) 2013

One way to implement the initial pass scheme is to provide "N-best" recognition results (Schwartz and Chow 1990; Murveit et al. 1993; Young 1993). The N-best output is simply a list of the top N most likely hypotheses, which can then be searched for a match.

The major disadvantage of N-best is its limited effectiveness – for $O(N)$ of 10–100 (Schwartz et al. 1992). The reason for this is that the resulting search process basically uses a finite number of available knowledge sources that constrain the space; the N-best choice and ordering is performed using only some of these sources, namely those that hold more constraining power at lesser cost. The rest of the sources are used only for searching the resulting N-best list (Schwartz and Austin 1991).

A more advanced "initial pass" implementation is the lattice (Murveit et al. 1993), which can effectively account for many more hypotheses. A lattice is a connected loop-free directed graph, constructed so that each node of the graph is associated with a point in time during the speech utterance, and labeled with a hypothesis (phoneme or word) and a score representing the likelihood of that hypothesis.

The lattice thus limits the space to all the likely recognition phoneme strings or word sequences, based on the technique used in the recognition pass. The actual search is then performed only on sequences contained in the lattice.

Different approaches have been suggested for implementing the lattice (Richardson et al. 1995). An off-line programming approach (James and Young 1994) can generate the phoneme lattices using the Viterbi algorithm. A modified variant of the Viterbi algorithm is based on the token passing paradigm (Young et al. 1989a). Similar methods have been employed in LVCSR research (Young 1993, 1996; Seide et al. 2004).

A variant of this approach is considering the search lattice as a phoneme-transition grammar where temporal and scoring information is left out, such that more hypotheses, which would have been eliminated by the scoring mechanism, remain implicit in the grammar and are included in the lattice (Murveit et al. 1993).

A leading method for reducing less probable paths during the search itself is the beam search and its variants (Haeb-Umbach and Ney 1994). In this data-driven approach (Ney et al. 1992), each sub-part of the input data (state, phoneme, word) is considered in relation to its preceding part and a probability is calculated as to where the new sub-part can construct a complete data part (phoneme, word, sentence). At the beginning of this process, many candidates are possible, which means that the search could considerably expand as it continues. To allow for a practical computation, the beam search drops unlikely combinations at each hypothesis calculation, according to parameters defined for the search, determined by a training process.

A further modification to the beam search is a tree organization of the lexicon (Haeb-Umbach and Ney 1994). Ney et al. showed a reduction of the search space by focusing on the initial phonemes of the words, which require most of the effort (Ney et al. 1992).

The scoring mechanism for building the lattice, and indeed also for the N-best choice, also varies between methods – from simple grammar rules ("legal" word sequences) (Schwartz et al. 1992; Demuynck et al. 2000) to Natural Language Processing (NLP) (Young et al. 1989b).

Another concept that may serve the same goal of search space reduction is anchor points. The term "anchor" has many connotations. In the search space notion, anchor points are positions within the search space around which the search is presumed to be more profitable. The space is therefore reduced to a set of much smaller sub-sets, each centered around an anchor point.

The use of anchor points in this context has been suggested in the field of pattern recognition (Ramasubramanian and Paliwal 1992). The common nearest-neighbor search, for example, aims to find the closest N point to a query point in K-dimensional space, where the complexity increases exponentially with K. In this application, all space vectors that lie outside a region that is defined around the anchor point are eliminated. The computation involved in this case is basically a spatial distance measure.

The concept of anchors can also be used in phonetic searching, where certain phonemes can serve as anchors, and the search for the keyword can be performed only around the anchor points (Tetariy et al. 2010, 2012). As described in Sect. 3.1, phonetic search KWS generates a search space (grid) G of size NxM which is linear to the number of keywords and length (in phonemes) of the speech database. The computation for each word entails calculating the distance measure between the word and every location on the grid. When searching very large databases (which is common in KWS applications), this is extremely cumbersome and may result in unacceptable response times. A search that is too time-consuming is not compatible with expectations for real-time or even near real-time responses and may render the application impractical for use.

In order for phonetic search KWS to be applicable in the real-world, the complexity of the process must be reduced to practical levels for searching very large speech databases (up to tens of thousands times faster than real time). This challenge is common to many areas of science and computer science, including machine learning (Srinivas and Patnaik 1991; Brin 1995), routing technologies (Guo and Matta 1999) and image searching (Moallem and Faez 2001).

4.2   Complexity Reduction in Phonetic Search

As with any search technology, the efficiency of a phonetic search is ascertained mainly by analyzing various aspects of processing time during the indexing and search stages, as well as, the accuracy of the search results. Ideally, the entire phoneme sequence representing a given keyword would be found within the searched database (also represented as a sequence of phonemes), thus resulting in a highly reliable keyword hypothesis. However, the reality is that phoneme recognition output inherently suffers from insertion, deletion and substitution errors, resulting in what is known as a high Phoneme Error Rate (PER). And as a result of these errors, the difference (in terms of phonemes) between the keyword transcription and the hypothesis is greater, thus negatively affecting its reliability. Furthermore, a lattice is often generated in order to provide multiple hypotheses and thus compensate for the high PER. This further increases the search space, and consequently, the computational complexity and response time.

Generally, a Minimum Edit Distance (MED) (Jurafsky 2000), also known as the Levenshtein distance, is used to calculate the minimum cost of transforming an hypothesized sequence of phonemes from the searched database into a sequence of phonemes representing a keyword. This calculation uses a combination of cost values for insertion, deletion and substitution. The MED is usually computed efficiently using a Dynamic Programming (DP) algorithm and thresholding of the MED score accepts or rejects a hypothesis and provides some robustness against phoneme recognizer errors (Thambiratnam 2005).

The need to reduce the computational complexity of a phonetic search has motivated several research studies. Basically, these explored the two steps involved in phonetic search; phoneme sequence (or lattice) generation and distance calculation. Since the phoneme recognition is performed only once, naturally most of the effort has been invested in various methods for reducing the search space. Within the search itself, most phonetic search methods use the edit distance thus leaving the main focus on reducing the number of hypotheses and respectively the number of edit distance calculations.

Dharanipragada and Roukos and Burget and Černocký et al. proposed methods for reducing the search space by "zooming" in on regions of speech where keywords are more probable to occur (Dharanipragada and Roukos 2002; Burget et al. 2006).

Thambiratnam and Sridharan proposed a Dynamic Match Lattice Spotting (DMLS) mechanism which is basically a phonetic search supplemented by beam pruning to reduce the search space (Thambiratnam and Sridharan 2007).

A different dynamic programming match approach on a path or utterance level was proposed by James and Young. Faster performance was achieved by labeling all keyword phonemes according to whether they are "weak" and thus can be deleted, inserted or substituted (according to a set of postulated constraining rules) or whether they are "strong" and thus must contain exact matches in the lattice in order that the construction of the keyword be pursued (James and Young 1994).

4.3   Anchor-Based Phonetic Search

This section will present a new algorithm for an efficient search that takes phonemes recognized with high reliability and utilizes them as anchors in the search process (Tetariy et al. 2010, 2012). The goal is to employ an anchor-based search alongside a classic edit distance in order to reduce the search space, and subsequently the computational complexity, while minimizing the effects on KWS recognition performance.

In order to illustrate the effectiveness of the anchor method, it should be compared to a search on an un-reduced search space; namely, an exhaustive search. In an exhaustive search, the matching process is performed across all possibilities in the space. That is, all phoneme sequences in the search space can be considered as hypotheses for a match to the searched word.

As described in Sect. 2.3, the phonetic search engine uses two types of input data; a list of keywords of size N, where each word is represented by a sequence of phonemes, and a speech database which has been run through a phoneme decoder to produce a length M sequence of recognized phonemes. Thus the exhaustive search process can be illustrated using the following example:

Given an input phoneme sequence of length M and a search for the keyword "NEWPORT," represented by its ARPAbet transcription {N UW P AO R T}; employing an exhaustive search on this keyword alone will necessitate the generation of M-5 hypotheses and a computation of M-5 distances, since all hypotheses would be generated and considered (see Fig. 7). A large speech database would result in a long M and require NxM grid point computations, the majority of which are unnecessary.

In order to produce all possible hypotheses for each keyword during an exhaustive search, it is generally necessary to calculate a large number of distance values. As in the basic notion of search reduction, many of these hypotheses are actually redundant or irrelevant since, realistically, most keywords are unlikely to occur more than a few times in a given database (Saraclar and Sproat 2004). By determining phoneme anchor points, it is possible to substantially decrease the number of hypotheses, and consequently, the number of distance calculations for each keyword within the search space (Tetariy et al. 2010). This leads to more focused and efficient hypothesis generation.

In the context of phonetic search KWS, an anchor point is a phoneme that is recognized by the phoneme decoder at a high level of reliability. Phoneme anchors can serve as focal points on the grid where the distance measure from a sequence

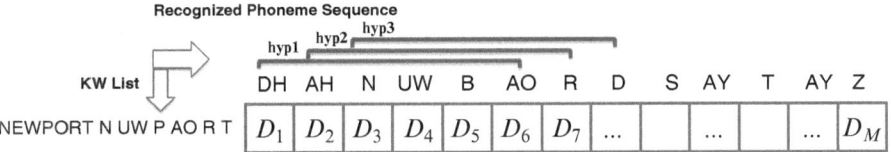

Fig. 7   An exhaustive search hypothesis

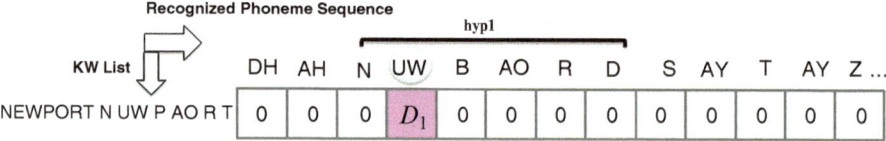

Fig. 8 A phoneme anchor point based hypothesis

around it and the keyword is sought. Since the anchor itself is reliably recognized, the computation is expected to yield a smaller distance and produce a match. The effectiveness of an anchor-based search process can be illustrated by the following example:

Given again the keyword "NEWPORT," represented by its transcription {N UW P AO R T}, and assuming the phoneme "UW" is known to be reliably recognized; then rather than computing the distance for all the M-5 hypotheses, as would be necessary in an exhaustive search, one distance calculation can be computed. This is demonstrated in Fig. 8. Thus, the correct selection of anchor points leads to a smaller number of hypotheses and consequently a significant reduction in computation time.

Using this method, the majority of the NxM grid points, that otherwise would have been considered in an exhaustive search, will not produce active hypotheses and thus will not be computed. The computational complexity of the phonetic search process using this anchor method is $O(kNM(Average(L))^2)$, where $Average(L)$ is the average length of the keywords in the vocabulary, and $0 < k < 1$ represents the ratio of the anchors to the recognized phoneme sequence. When the value of k is small, computational complexity is significantly lower than in the exhaustive search.

The challenge, of course, lies in determining which phonemes can be used as anchors. The criterion for this choice is that the phonemes should be reliably recognized. Reliability is, as usual, relative, and therefore should be defined according to a certain threshold. To achieve this, a number of mechanisms for anchor point selection have been examined. These mechanisms can be divided into two main types: a priori selection of phoneme anchor points; and real-time, dynamic selection of phoneme anchor points (Tetariy et al. 2010).

A priori anchor selection can be further divided into anchor selection according to language frequency (phonemes that can be found in a statistically large number of words); and anchor selection according to high phoneme recognition rates by the phoneme decoder. The advantage of selecting anchor points in an a priori manner is that it is a one-time process that ends before the search begins and does not expend computational resources during the phonetic search process itself. The disadvantage is that not every input phoneme sequence or keyword may contain the anchors selected. In these cases, it would be necessary to execute a full exhaustive search on all possible search space points in the relevant phoneme sequence, making the overall search space reduction less efficient.

The real-time dynamic phoneme anchor selection method was designed to resolve this. This method utilizes relatively reliable phonemes that are common to both the keyword and the database phoneme sequences and are determined in

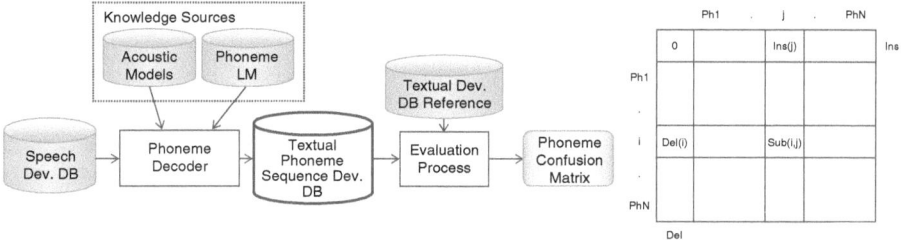

Fig. 9 Confusion matrix creation process resulting in a 3-part matrix

real-time during the search process itself. This leads to a dynamic procedure where a different anchor is determined for every searched keyword. Although this adds some computation to the search process itself, it also contributes to a major decrease in the number of hypotheses.

The reliability of anchor phoneme selection can be determined by analyzing the phoneme decoder's confusion matrix. A confusion matrix is often used for similarity measure in speech applications (Doddington 1989; Witbrock and Hauptmann 1997; Bouselmi et al. 2005; Žgank et al. 2005) as it shows the nature of the recognizer's mistakes. The result is a 3-part matrix, highlighting phoneme substitutions, insertions and deletions. The confusion matrix creation process is illustrated in Fig. 9.

The matrix contains the reference and recognized phonemes in the rows and columns $(Ph_{i\ and}\ Ph_j)$ respectively, and each cell contains the number of times a particular phoneme was substituted by another, deleted, inserted or correctly recognized. The confusion matrix elements C(i,j) are defined as follows:

1. $C(i,0)$ – is the number of times that Ph_i was deleted at the output of the engine.
2. $C(0,j)$ – is the number of times that Ph_j was inserted at the output of the engine.
3. $C(i,j)$ – is the number of times that the input of the engine was Ph_i which was replaced by Ph_j at the engine output (when Ph_i is equal to Ph_j, Ph_i was correctly recognized).

Based on these elements, we can define the probabilities of correct, substitution, insertion and deletion for each phoneme (PN – the number of phonemes in the phoneme set) as follows:

$$
\begin{cases}
P_{Sub}(i,j)/Corr(if\ i = j) = \dfrac{C(i,j)}{\sum_{k=1}^{k=PN} C(k,j)} & for \quad 1 \leq i,j \leq PN \\[3mm]
P_{Ins(j)} = \dfrac{C(0,j)}{\sum_{k=1}^{k=PN} C(k,j)} & for \quad 1 \leq j \leq PN \\[3mm]
P_{del(j)} = \dfrac{C(j,0)}{\sum_{k=1}^{k=PN} C(j,k)} & for \quad 1 \leq j \leq PN
\end{cases}
\tag{4.1}
$$

These probabilities can then be used to define a posterior quality measure per phoneme, which is a weighted sum of all the posterior probabilities, as follows:

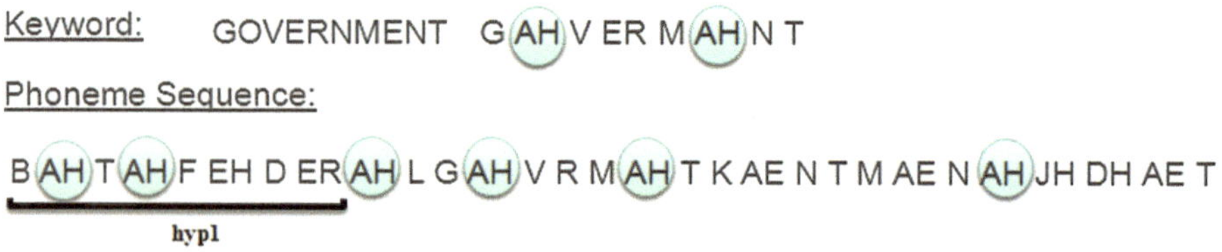

Keyword: GOVERNMENT   G AH V ER M AH N T

Phoneme Sequence:

B AH T AH F EH D ER AH L G AH V R M AH T K AE N T M AE N AH JH DH AE T

hyp1

hyp1 - BAH T AH F EH D ER	hyp6 - D ER AH L G AH V R
hyp2 – T AH F EH D ER AH L	hyp7 - M AH T K AE N T M
hyp3 - ER AH L G AH V R M	hyp8 - G AH V R M AH T K
hyp4 – AH F EH D ER AH L G	hyp9 - N T M AE N AH JH DH
hyp5 – G AH V R M AH T K	

Fig. 10 Hypothesis generation using real-time dynamic anchors

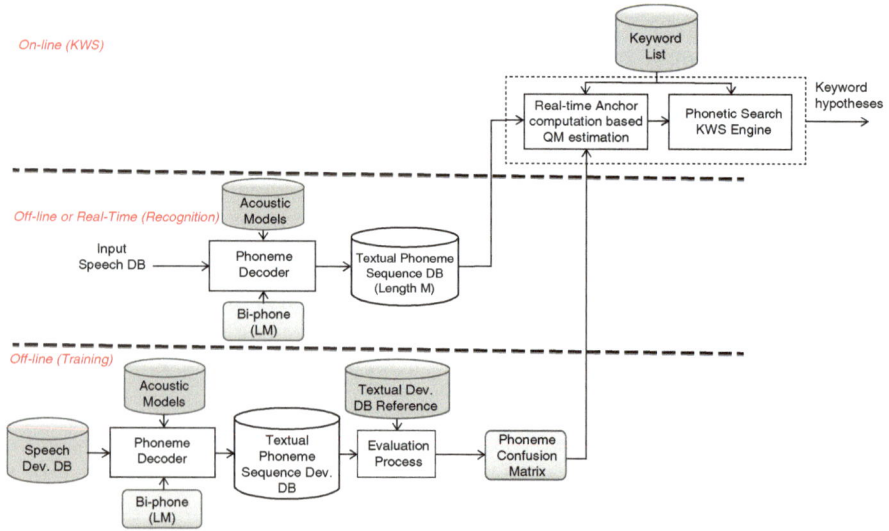

Fig. 11 One-time transformation of a speech DB to a textual phoneme DB and on-line KWS phonetic search engine using a real-time dynamic anchor-based search

$$QM_{Phoneme(k)} = \alpha_1 \cdot P_{k,C} - \alpha_2 \cdot P_{k,S} - \alpha_3 \cdot P_{k,I} \qquad (4.2)$$

where α_i are weighting coefficients determined based on empirical testing and $P_{k,C}$, $P_{k,S}$ and $P_{k,I}$ are the respective correct, substitution and insertion probabilities for the kth phoneme.

Anchor selection is performed by choosing only the phonemes whose QM exceeds a given threshold. The threshold can be set based on performance results. This allows new anchor points to be selected for each keyword, based on the new input phoneme sequence.

Figure 10 demonstrates the hypothesis generation process, where all of the phonemes representing the keyword can be found in the recognized phoneme sequence; but, the anchor phoneme selected by the QM based on the chosen threshold was AH. In this case, only nine hypotheses were computed instead of 21.

The final framework of the suggested keyword spotting system, which consists of a combination of real-time and offline processing, is described in Fig. 11.

Chapter 5
Evaluating Phonetic Search KWS

Complexity reduction algorithm evaluation should be carefully performed to assess its performance and usability. A basic evaluation in such cases measures two aspects in comparison to the exhaustive search: the relative decrease in computational complexity and the relative change (increase or decrease) in recognition performance in the reduced computational complexity mode.

In addition, evaluation of the phonetic search itself is performed using textual data which is transformed into a sequence of phonemes simulating 100% phoneme recognition.

The next section describes the performance metrics, evaluation process and evaluation databases while the following section describes the evaluation results.

5.1 Performance Metrics

The success of a keyword spotting task is measured on the basis of two main metrics: the percentage of words correctly recognized, referred to as the Detection Rate (DR), and on the number of false alarms per hour per vocabulary word, referred to as the False Alarm Rate (FAR).

To better explain these metrics, consider a speech signal (or DB) and a given KWS vocabulary and define:

$W = \{W_i\}$	The sequence of words describing the content of the speech signal
DS	The duration in hours of the speech signal
KWS-VOC-SIZE	The size of the KWS vocabulary
KW-NUM	The number of actual keywords included in the speech signal (with all its repetitions)

A. Moyal et al., *Phonetic Search Methods for Large Speech Databases*, SpringerBriefs in Speech Technology, DOI 10.1007/978-1-4614-6489-1_5, © The Author(s) 2013

Now, consider the output of a KWS engine which is a list of recognized keywords and define:

KW-OUT	The total number of keywords detected by the engine
KW-OUT-CORRECT	The number of keywords in the speech signal correctly recognized
KW-OUT-NOT	The number of keywords in the speech signal that were not recognized
KW-OUT-FA	The number of keywords in the output that represent false alarms

Based on these, the following metrics can be defined:

Detection Rate	$DR = (KW\text{-}OUT\text{-}CORRECT/KW\text{-}NUM)*100$
Missed Detection	$MD = (KW\text{-}OUT\text{-}NOT/KW\text{-}NUM)*100$
False Alarm	$FA = (KW\text{-}OUT\text{-}FA/KW\text{-}NUM)*100$
False Alarm Rate	$FAR = KW\text{-}OUT\text{-}FA/(DS*KWS\text{-}VOC\text{-}SIZE)$

The DR and FAR values are used to determine the working point for the KWS engine. The working point performance is usually based on a distance threshold that controls the tradeoff between the DR and FAR. The KWS engine generates KWS hypotheses, each with a distance value. A threshold for the distance value allows the system to determine whether a given keyword hypothesis will be labeled as a spotted word or rejected. Plotting the DR and FAR values as a function of the distance level on a graph enables an analysis of the KWS engine performance and facilitates selecting the optimal working point for a given application.

A sample graph of DR and FAR values as a function of the distance threshold is presented in Fig. 12:

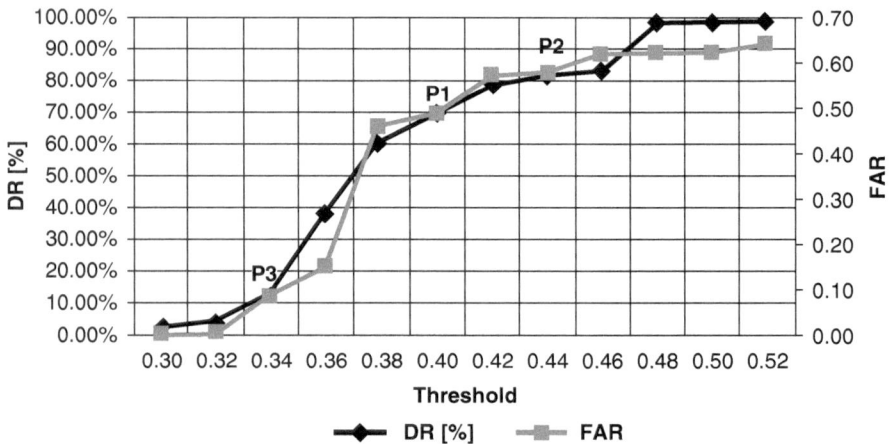

Fig. 12 FAR and DR as a function of threshold

The graph has three sample points marked as P1, P2 and P3, representing several working points with different combinations of DR and FAR based on three different distance thresholds.

The P1 values resulted from a distance threshold of 0.4 which led to a DR of 70% and a FAR of 0.5 words per hour per vocabulary word. Increasing the distance threshold from 0.4 to 0.44 produces the P2 values of an 81% DR and a FAR of 0.58, showing that increasing the threshold results in the rejection of fewer hypotheses, thus resulting in higher DR and FAR values.

On the other hand, decreasing the distance threshold leads to the rejection of a larger number of hypotheses, thus resulting in lower DR and FAR values. When the threshold is reduced from 0.4 to 0.34, the resulting P3 values have a DR of 13% and a FAR of 0.1.

KWS performance is also commonly represented by a Receiver Operating Characteristic (ROC) curve, which presents the DR as a function of FAR with the distance threshold as a parameter. The same data presented above is shown in Fig. 13 as a ROC curve.

It is important to understand that FAR is an engine characteristic in the sense that even if there are no keywords in the speech input, the KWS engine still produces false alarms. The DR, on the other hand, reflects the capability of the KWS engine to correctly recognize words that can only be seen when the input contains words from the searched vocabulary.

Take, for example, a KWS engine working on 2 hours of speech with a DR = 50%, a FAR of 2, and a KWS vocabulary of 50 words, where the first hour contains 200 words from the searched vocabulary and the second hour contains no words from the searched vocabulary. The KWS system output for the first hour of speech will be 200 alerts: 100 correctly recognized words and 100 false alarms. The KWS system output for the second hour of speech will be 100, all of which are false alarms.

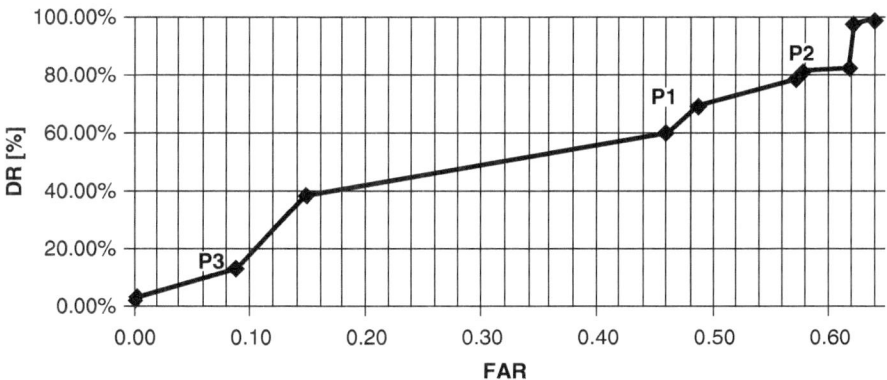

Fig. 13 ROC curve representing the DR as a function of FAR

5.2 Evaluation Process

The evaluation process of a KWS engine requires:

1. A representative transcribed speech database.
2. Time segmentation for the KWS vocabulary in the speech database.
3. A KWS engine with distance threshold flexibility.
4. A KWS evaluation tool that computes the DR and FAR for varying distance thresholds. A minimal time overlap threshold should be defined in order to define a hypothesis as correct or false.

The evaluation process should begin by operating the KWS engine on the speech database with no threshold. This will generate all possible hypotheses, each with its corresponding distance score. At the second stage of the evaluation process, the previous output should be used to compute the DR and FAR for various distance thresholds, thus producing a global graph of DR and FAR values as a function of the distance threshold.

This graph will enable the selection of a working point which, once decided upon, will determine the distance threshold that should be set for the KWS engine operation.

The evaluation speech database should be representative of a target database used in operating the KWS application with respect to the following parameters:

1. Number of speakers
2. Speaker background: gender, age and dialect
3. Recording environments and channels
4. The ability to define several KWS vocabularies
5. Wealth of spoken words beyond the KWS vocabulary

The evaluation speech database should be fully transcribed and the intended keywords should also include time segmentation, which is crucial for a thorough evaluation. For example, in a database with no segmentation, a keyword recognized within a given input sequence of phonemes that indeed contains that word, would be considered correctly recognized regardless of where the keyword actually occurred. However, if the database is time segmented, the recognized keyword must overlap with the location where the same word was marked in the database, otherwise the recognition should actually be considered a false alarm.

A minimal time overlap threshold should be defined with regard to segmentation points of the source and recognized words. The full transcription is required for error analysis.

Some applications require locating speech files that contain a certain word or words for classification or later analysis. In these cases, the specific time stamp of a keyword may be less crucial, and thus the tedious and time-consuming process of time segmentation may be superfluous. The evaluation results in these cases are expected to be higher or equal to strict evaluation based on time stamps.

Table 1 Summary of VM and Mac databases

	VM	Mac
Unique words	7,081	2,130
Number of utterances	2,483	736
Ave. utterance duration (s)	22.30	3.55
Ave. number of words per utterance	65.97	9.07
Ave. phoneme sequence length at phoneme decoder output per utterance	242	38

It is recommended to use two similarly defined databases, one for development and one for evaluation. The development database should be used for the various experiments needed to set the distance threshold; the evaluation database should be used for assessing KWS performance.

5.3 Evaluation Databases

Three speech databases (DBs) were used to evaluate the suggested anchor-based method: IBM Voicemail I (Padmanabhan et al. 1998) and IBM Voicemail II (Padmanabhan et al. 2002), together referred to as VM, which consist of recorded spontaneous voice messages; and the Wall Street Journal portion of the Macrophone DB (Mac) (Bernstein et al. 1994), which contains a collection of read sentences. The characteristics of the databases are summarized in Table 1.

Chapter 6
Evaluation Results

In order to evaluate the performance of the suggested anchor-based algorithm in reducing the computational complexity, the size of the search space and the KWS performance were calculated for the anchor-based search and compared to an exhaustive search. The reduction in computational complexity was determined by measuring the average runtime for processing an input phoneme string and then retrieving the keywords using a standard server. The relative decrease in runtime was measured rather than absolute figures (which may change depending on the server used).

In order to define a list of keywords, the vocabulary words found in each DB were filtered based on a number of criteria meant to ensure that the keywords are relatively common and not too short: a minimum frequency relative to size of the DB; a minimum transcription length of five phonemes; a noun or verb part of speech. Once the lists were filtered, the top 50 words were selected as keywords.

The expected computational complexity of the exhaustive search and the suggested algorithm using the parameters of the DBs is presented in Table 2. The tests were run on an input sequence of phonemes, but could also be run on a lattice. The results show that the average processing time for searching the keyword list using the suggested anchor-based algorithm on VM recordings was 2.41 seconds compared to 19.41 seconds required by the exhaustive search, when the average duration of a VM utterance was 22.3 seconds. For Mac, the average processing time was 0.15 seconds compared to the 1.16 seconds required by exhaustive search, when the average duration of a Mac utterance was 3.55 seconds. A reduction of almost 90% in processing time was achieved.

6.1 Exhaustive Search

In order to test the effectiveness of the overall mechanism, including the distance measure and the threshold-based decision, a synthetic experiment was first conducted as a benchmark. The FAR and DR were measured as a function of the

A. Moyal et al., *Phonetic Search Methods for Large Speech Databases*, SpringerBriefs in Speech Technology, DOI 10.1007/978-1-4614-6489-1_6, © The Author(s) 2013

Table 2 Computational complexity reduction

Search type	Exhaustive		Anchor-based	
Computational complexity	$O(NM(Average(L))^2)$		$O(kNM(Average(L))^2)$	
DB	VM		Mac	
Search type	Exhaustive	Anchor-based	Exhaustive	Anchor-based
Processing time (ave. sec)	19.41	2.41	1.16	0.15
Processing time reduction (%)	–	87.58%	–	86.92%
Search space reduction (%)	–	90.58%	–	86.09%

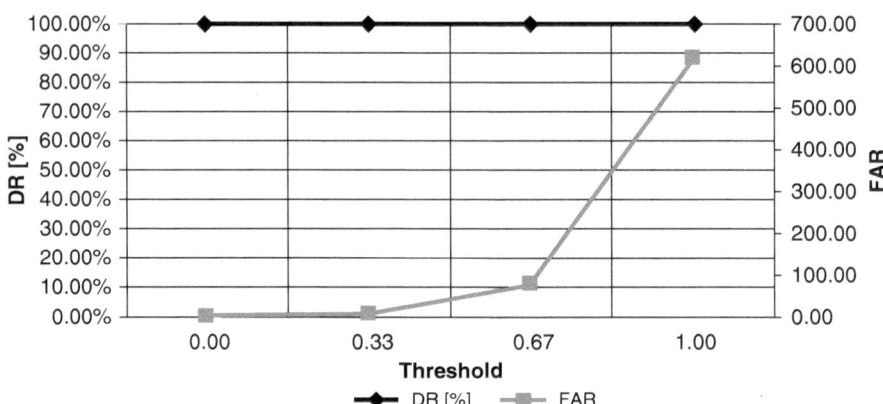

Fig. 14 FAR and DR as a function of threshold using 100% phoneme recognition from the VM DB

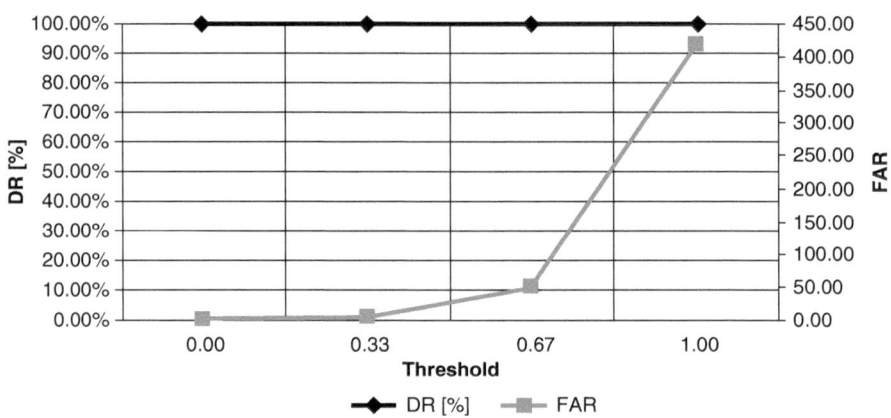

Fig. 15 FAR and DR as a function of threshold using 100% phoneme recognition from the Mac DB

threshold assuming 100% phoneme recognition (that is, rather than using the phoneme sequences resulting from the phoneme decoder, the orthographic transcription of the speech utterances was converted to a phoneme string by using the lexicon of ARPAbet transcriptions). The same experiment was then conducted on the actual output of a phoneme decoder and the results were compared.

6.1.1 Textual Benchmark

The distribution of the FAR and DR, as a function of the threshold for this experiment, is presented in Figs. 14 and 15 for the VM and Mac databases, respectively.

The graphs show that already at a threshold of zero, it is possible to reach a DR of 100% (exact match), while the FAR increases with respect to the increase of the threshold.

6.1.2 KWS on Speech

6.1.2.1 Single Threshold

As stated in Sect. 3.1, performing an exhaustive search entails the construction of all possible hypotheses for matching, such that each keyword can be considered as possibly occurring on any location of the input sequence of phonemes.

The distribution of the FAR and DR, as a function of the threshold for this experiment, is presented in Figs. 16 and 17 for the VM and Mac databases, respectively.

The graph shows several potential working points plotted for each database. The working point selected is highly dependent on the importance that the KWS-based application user assigns to the DR and the FAR. As a representative example, a threshold of 1.0 was selected. The results can be seen in Table 3.

6.1.2.2 Multiple Thresholds

An analysis of the distance values per word showed a different dynamic range for each word and led to the conclusion that using a different threshold for each word may lead to better performance. Thus, in the following evaluation, instead of a single threshold for all keywords, a separate threshold was calculated for each individual keyword. The thresholds were calculated for each word based on the highest DR with FAR values no higher than an absolute value that can be set by the user (in our case, 5).

Table 4 demonstrates the average and Standard Deviation (SD) of the threshold, FAR and DR for the two databases.

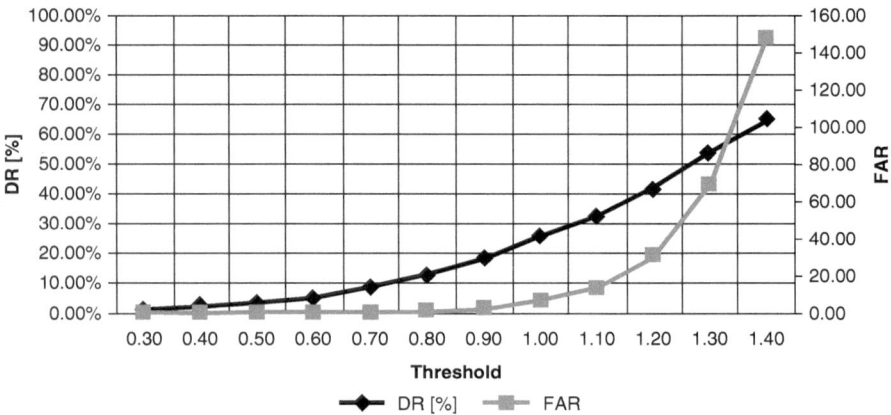

Fig. 16 FAR and DR as a function of threshold in an exhaustive search using actual phoneme recognition results from the VM DB

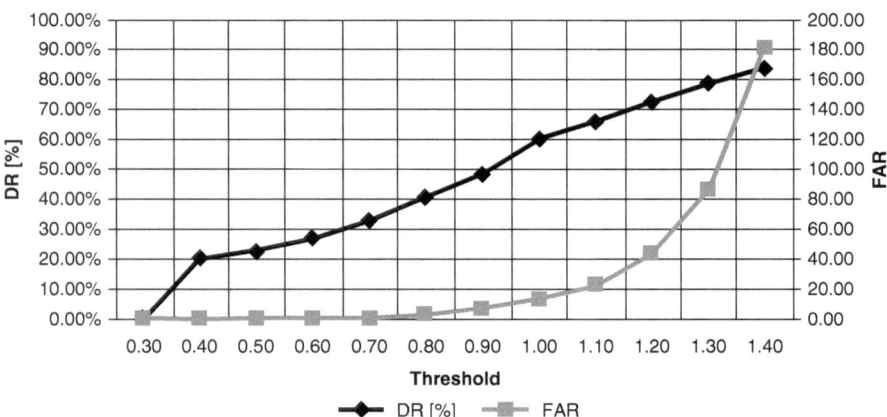

Fig. 17 FAR and DR as a function of threshold in an exhaustive search using actual phoneme recognition results from the Mac DB

Table 3 A single threshold for all keywords in an exhaustive search

	VM	MAC
Threshold	**1.0**	**1.0**
FAR	6.63	12.71
DR	25.34%	59.85%

Table 4 KWS performance using a threshold per word based on an exhaustive search

	VM		Mac	
	Average	SD	Average	SD
Threshold	**1.47**	**0.14**	**1.37**	**0.23**
FAR	3.13	0.07	2.45	1.79
DR	79.05%	10.51	85.50%	19.10

6.2   Anchor-Based Search

In order to test the effectiveness of the overall mechanism, including the distance measure and the threshold-based decision, a similar evaluation was conducted using the suggested anchor-based search. Again, first a benchmark synthetic experiment was performed, where the FAR and DR were measured as a function of the threshold assuming 100% phoneme recognition. The assumed 100% phoneme accuracy in this experiment makes the QM irrelevant. Thus the algorithm used in the synthetic experiment treats anchors as phonemes that are common to the keyword and the recognized phoneme string, where no more than two anchors are selected per word. The same experiment was then conducted on the actual output of a phoneme decoder and the results were compared.

6.2.1   Textual Benchmark

The distribution of the FAR and DR, as a function of a single threshold for all keywords in the synthetic experiment, is presented in Figs. 18 and 19 for the VM and Mac databases, respectively.

As in the case of the exhaustive search, the graphs show that already at a threshold of zero, it is possible to reach a DR of 100% (exact match), while the FAR increases with respect to the increase of the threshold. However, in the anchor-based search the FAR values are lower as a result of the search space reduction (fewer hypotheses).

6.2.2   Reduced Complexity KWS on Speech

6.2.2.1   Single Threshold

The same experiment was repeated using actual phoneme recognition results on the same databases.

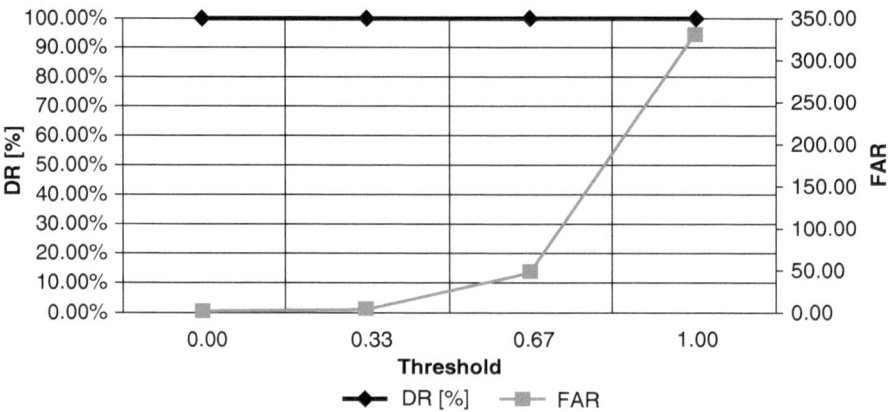

Fig. 18 FAR and DR as a function of threshold in an anchor-based search assuming 100% phoneme recognition results from the VM DB

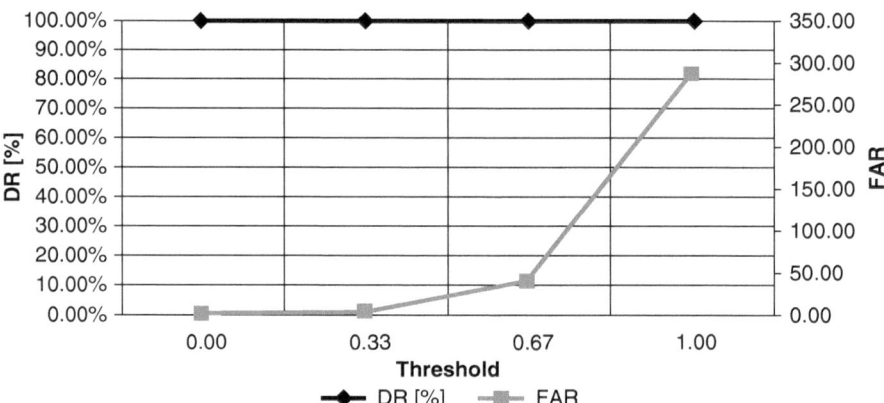

Fig. 19 FAR and DR as a function of threshold in an anchor-based search assuming 100% phoneme recognition results from the Mac DB

The distribution of the FAR and DR, as a function of the threshold for this experiment, is presented in Figs. 20 and 21 for the VM and Mac databases, respectively.

The graph shows several potential working points plotted for each database. The working point selected is highly dependent on the importance that the KWS-based application user assigns to the DR and the FAR.

As a representative example, a threshold of 1.0 was selected, leading to similar DR values for both DBs, but with a significant variation in the FAR. The results are shown in Table 5.

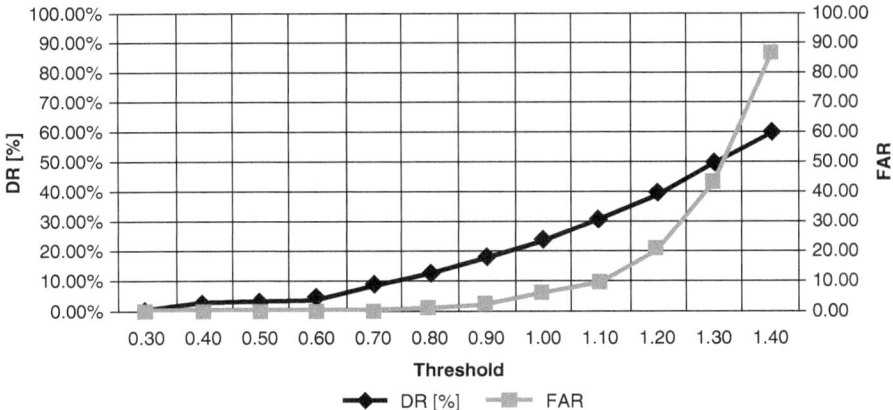

Fig. 20 FAR and DR as a function of threshold in an anchor-based search using actual phoneme recognition results from the VM DB

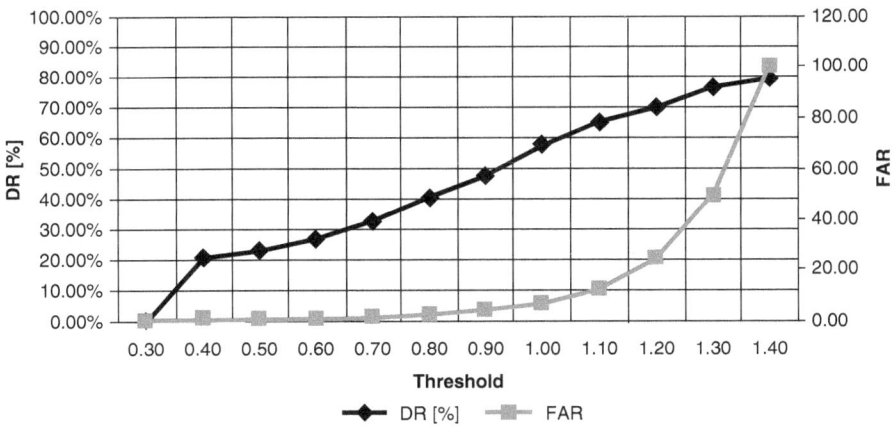

Fig. 21 FAR and DR as a function of threshold in an anchor-based search using actual phoneme recognition results from the Mac DB

Table 5 A single threshold for all keywords based on anchor points compared to an exhaustive search

		VM	Mac
Threshold		**1.0**	**1.0**
FAR	**Anchor**	**4.59**	**6.30**
	Exhaustive	6.63	12.71
DR	**Anchor**	**23.65%**	**57.95%**
	Exhaustive	25.34%	59.85%

The results indicate that using the anchor-based approach in KWS reduces the size of the search space and the processing time by 86–90% in comparison to an exhaustive search, while at the same time showing a 31% relative reduction in the FAR for the VM database and a 50% relative reduction in the FAR for Mac. In both cases, the DR was only minimally affected; with a 7% and 3% relative decrease respectively.

The anchor point algorithm optimizes the hypothesis generation process in a phonetic search. This is because elements on the grid are rarely explored in vain. A correct selection of anchor points leads to a smaller number of hypotheses resulting in a search with a rapid execution time.

6.2.3   Multiple Thresholds

Similar to the exhaustive search experiment, analysis of the results indicates that the distance values vary in dynamic range between words. Again, the conclusion was to use a different threshold for different words in order to improve performance. Thus in the second evaluation, instead of a single threshold for all keywords, a separate threshold was calculated for each individual keyword. The thresholds were calculated for each word based on the highest DR with a FAR no higher than an absolute value that can be set by the user (in our case, 5). The performance of the KWS task in an anchor-based search can be measured by comparing the average and SD of the threshold, the FAR and the DR for the two databases to the same parameters from the exhaustive search.

Table 6 shows that the KWS performance in the anchor-based search is no worse than an exhaustive search using a threshold per word. With an almost identical threshold, there is virtually no change in the FAR for either the VM or Mac DBs. With respect to the DR, there is even an increase in performance, with the VM DB showing a relative decrease of 20% in the error rate and the Mac DB showing a decrease of 7% (a 4% improvement in average error rate from the exhaustive search in VM and a 1% improvement in Mac).

Table 6 KWS performance using a threshold per word based on anchor points compared to exhaustive search

		VM		Mac	
		Average	SD	Average	SD
Threshold	**Anchor**	**1.57**	**0.16**	**1.3**	**0.21**
	Exhaustive	1.47	0.14	1.37	0.23
FAR	**Anchor**	**3.11**	**0.06**	**2.49**	**1.7**
	Exhaustive	3.13	0.07	2.45	1.79
DR	**Anchor**	**83.16%**	**7.98**	**86.51%**	**18.75**
	Exhaustive	79.05%	10.51	85.50%	19.10

6.3 Lessons Learned from the Evaluation

The phonetic search process is hampered by high computational complexity when performed on large speech DBs, particularly when searching within a phoneme lattice. However, the evaluation findings show that the anchor-based search algorithm can substantially reduce this computational complexity, making the phonetic search process usable for applications demanding rapid search capabilities in large speech DBs.

Prior to beginning the evaluation on real speech, a synthetic bench-mark experiment was performed in order to validate the quality of the KWS mechanism. The synthetic experiment used a phoneme string that simulated 100% phoneme recognition accuracy. The results showed that the KWS mechanism produces a 100% detection rate with a relatively low false alarm rate.

The evaluation of actual speech showed that the suggested anchor-based algorithm reduces the size of the search space and the processing time by 86–90% in comparison to an exhaustive search, while at the same time showing a 31–50% relative reduction in the FAR and a 3–7% relative decrease in the DR.

Performing an exhaustive search with different distance thresholds for each of the keywords, dramatically improves the DR in comparison to using a single threshold (VM: 25.34% Mac: 59.85% vs. VM: 79.05% and Mac: 85.50%), while also contributing to a decrease in the FAR values (VM: 6.63 Mac: 12.71 vs. VM: 3.13 and Mac: 2.45).

By implementing the anchor-based search algorithm, the results improved when using a single threshold for all keywords and when using a separate threshold for each keyword. In the case of multiple keywords, the FAR values remained at the same level while the DR improved slightly.

When a single threshold was used, there was a large discrepancy between the Mac and VM results, but the DR in both cases was only minimally affected. This is most likely due to the fact that the Mac DB consists of read speech and the VM DB consists of spontaneous speech. Spontaneous speech is known to consist of longer utterances, a large number of speech disfluencies and background noise. These characteristics lead to much poorer phoneme recognition results, making this seeming inconsistency less surprising.

When a separate threshold was used for each keyword, the DR drastically improved, thus producing similar KWS performance for both spontaneous and read speech.

Evidently, the distance threshold value has a major influence on the working point of the system in the sense of the trade-off between the FAR and DR rates. Further analysis suggests that the dynamic range of the values differs per word, and thus, a different threshold per word can dramatically improve performance.

Chapter 7
Summary

Speech recognition technology can be used for a wide range of applications. Keyword spotting is one of the more practical implementations of speech recognition, as it does not require any understanding of the transcribed speech, nor does it necessarily demand full transcription accuracy.

Naturally, the leading approach in ASR technology, LVCSR, is often translated to other speech processing domains and KWS is no exception. However, because KWS is generally performed on extremely large speech databases, LVCSR is not necessarily the most practical method. Fully transcribing huge amounts of speech is a computationally complex process that demands knowledge sources such as a very large recognition vocabulary and complex language model that bind the outcome transcription to decisions that are difficult to reverse. The level of complexity and lack of flexibility found in LVCSR KWS mechanisms, coupled with the parallel demands of application users for vocabulary-independent and fast KWS, have led researchers to search for alternate solutions.

Phonetic search is one method that researchers have turned to. Like the LVCSR method, phonetic search transforms the speech into text prior to beginning the KWS task. This is an advantage, as the transformation is an off-line one-time process, after which KWS spotting can be performed quickly and repeatedly. However, unlike LVCSR KWS, the resulting text is not an attempt at transcribing the speech word for word, but rather is a low-level transcription at the phonemic level. Although, the phonemic transcription may also be laden with errors, these can be overcome by generating phoneme lattices representing multiple hypotheses and by using smart distance calculations that compare the keyword transcriptions with the database transcription. This means that new keywords can be searched for without ever having to rerun the textual transformation stage before a search.

Still, however, the phonetic search KWS process is hampered by high complexity when performed on large speech databases; a situation unacceptable for real-world applications. Various methods have been suggested to reduce the computational complexity of the search. Some have aimed for accelerating the search itself, while others at better organizing the searched database through efficient indexing, in order to optimize it for quick retrieval.

A. Moyal et al., *Phonetic Search Methods for Large Speech Databases*, SpringerBriefs in Speech Technology, DOI 10.1007/978-1-4614-6489-1_7, © The Author(s) 2013

This brief suggests an anchor-based search algorithm that reduces this computational complexity and makes the phonetic search process usable for applications needing rapid searching on large speech DBs. The analysis addresses the phonetic search in its generic form by focusing on the selection of reliable hypotheses.

A reduction of almost 90% in search space and computational complexity of phonetic search KWS was achieved by using a phoneme anchor point based search algorithm.

Prior to beginning the experiment on real speech, a synthetic experiment was designed to validate the quality of the KWS mechanism. The synthetic experiment used a phoneme string that simulated 100% phoneme recognition accuracy. The results showed that the KWS mechanism produces a 100% detection rate with a relatively low false alarm rate.

The results of the KWS using actual speech show that it is possible to achieve an efficient phonetic search KWS method with a 90% reduction in search space and consequently runtime, while even improving the FAR and with some decrease in the detection rate, in comparison to the exhaustive search.

Not directly connected to the efficient search itself, but a key feature of the KWS algorithm, is the value of the distance threshold. The threshold determines the KWS system working point and controls the trade-off between the FAR and DR rates. Further analysis of the distance value shows that the dynamic range of the values differs per word, and thus, a distinct threshold for each keyword can dramatically improve performance. This is valid in both search methods – exhaustive and anchor-based.

Glossary of Acronyms

ASR	Automatic Speech Recognition
DB	Database
DP	Dynamic Programming
DR	Detection Rate
FAR	False Alarm Rate
HMI	Human Machine Interaction
HMM	Hidden Markov Model
KWS	Keyword Spotting
LM	Language Model
LVCSR	Large Vocabulary Continuous Speech Recognition
MED	Minimum Edit Distance
NLP	Natural Language Processing
OOV	Out of Vocabulary
PER	Phoneme Error Rate
ROC	Receiver Operating Characteristic
SD	Standard Deviation
SNR	Signal to Noise Ratio

A. Moyal et al., *Phonetic Search Methods for Large Speech Databases*, SpringerBriefs in Speech Technology, DOI 10.1007/978-1-4614-6489-1, © The Author(s) 2013

References

Alon G (2005) Key-word spotting – the base technology for speech analytics. Rishon lezion, NSC – natural speech communications

Amir A, Efrat A et al (2001) Advances in phonetic word spotting. In: Tenth international conference on information and knowledge management, Atlanta

Baker J, Deng L et al (2009) Developments and directions in speech recognition and understanding, Part 1 [DSP Education]. Signal Process Mag IEEE 26(3):75–80

Barras C, Allauzen A et al (2002) Transcribing audio-video archives. In: 2002 I.E. international conference on acoustics, speech, and signal processing (ICASSP). IEEE, Orlando

Bar-Yosef Y, Aloni-Lavi R et al (2012) Cross-language phonetic search for keyword spotting. In: Proceedings of 2012 speech processing conference, Tel-Aviv

Bernstein J, Taussig K et al (1994) Macrophone. Linguistic data consortium (LDC), Philadelphia

Bingham E, Mannila H (2001) Random projection in dimensionality reduction: applications to image and text data. In: The seventh ACM SIGKDD international conference on knowledge discovery and data mining, ACM, San Francisco

Bouselmi G, Fohr D et al (2005) Fully automated non-native speech recognition using confusion-based acoustic model integration. In: Interspeech, pp 1369–1372

Brin S (1995) Near neighbor search in large metric spaces. In: 21st International conference on very large data bases, VLDB '95, Zurich, September, pp 574–584

Burget L, Černocký J et al (2006) Indexing and search methods for spoken documents. In: Text, speech and dialogue 4188/2006 of Lecture notes in computer science. Springer, Berlin/Heidelberg, pp 351–358

Butzberger J, Murveit H et al (1992) Spontaneous speech effects in large vocabulary speech recognition applications. In: Workshop on speech and natural language, Association for Computational Linguistics

Cardillo PS, Clements M et al (2002) Phonetic searching vs. LVCSR: how to find what you really want in audio archives. Int J Speech Technol 5(1):9–22

Clements M, Cardillo PS et al (2001) Phonetic searching of digital audio. In: Proceedings of the broadcast engineering conference, Washington

Demuynck K, Duchateau J et al (2000) An efficient search space representation for large vocabulary continuous speech recognition. Speech Commun 30(1):37–53

Deng L, Huang X (2004) Challenges in adopting speech recognition. Commun ACM 47(1):69–75

Dharanipragada S, Roukos S (2002) A multistage algorithm for spotting new words in speech. IEEE Trans Speech Audio Process 10(8):542–550

Doddington GR (1989) Phonetically sensitive discriminants for improved speech recognition. In: International conference on acoustics, speech, and signal processing (ICASSP'89), IEEE, Glasgow

A. Moyal et al., *Phonetic Search Methods for Large Speech Databases*, SpringerBriefs in Speech Technology, DOI 10.1007/978-1-4614-6489-1, © The Author(s) 2013

Dubois C, Charlet D (2008) Using textual information from LVCSR transcripts for phonetic-based spoken term detection. In: IEEE international conference on acoustics, speech and signal processing (ICASSP'08), Las Vegas

Evermann G, Chan H et al (2005) Training LVCSR systems on thousands of hours of data. In: IEEE ICASSP, 2005, vol 1, pp 209–212

Furui S (2003) Recent advances in spontaneous speech recognition and understanding. ISCA & IEEE workshop on spontaneous speech processing and recognition

Furui S, Deng L et al (2012) Fundamental technologies in modern speech recognition. IEEE Signal Process Mag (IEEE Signal Processing Society) 26:16–17

Gishri M, Silber-Varod V (2010) Lexicon design for transcription of spontaneous voice messages. In: Seventh conference on international language resources and evaluation (LREC'10), Valetta

Gosztolya G, Tóth L (2011) Spoken term detection based on the most probable phoneme sequence. In: 2011 I.E. 9th international symposium on applied machine intelligence and informatics (SAMI), IEEE, Smolenice

Guo L, Matta M (1999) Search space reduction in QoS routing. In: 19th IEEE international conference on distributed computing systems. IEEE, Austin, pp 142–149

Gusfield D (1997) Algorithms on strings, trees and sequences: computer science and computational biology. Cambridge University Press, Cambridge

Haeb-Umbach R, Ney H (1994) Improvements in beam search for 10000-word continuous-speech recognition. IEEE Trans Speech Audio Process 2(2):353–356

Har-Lev B, Aharonson V et al (2010) An efficient phoneme distance measure using a lexical tree. In: IEEE 26th convention of electrical and electronics engineering in Israel, IEEE, Eilat

Heigold G, Nguyen P et al (2012) Investigations on exemplar-based features for speech recognition towards thousands of hours of unsupervised, noisy data. In: IEEE international conference on acoustics, speech and signal processing (ICASSP), IEEE, Kyoto

Hermelin D, Landau GM et al (2009) A unified algorithm for accelerating edit-distance computation via text compression. In: 26th international symposium on theoretical aspects of computer science, Feiburg

Hirsch HG, Pearce D (2000) The Aurora experimental framework for the performance evaluation of speech recognition systems under noisy conditions. In: ASR2000-automatic speech recognition: challenges for the new millennium ISCA tutorial and research workshop (ITRW)

Huo Q, Jiang H et al (1997) A Bayesian predictive classification approach to robust speech recognition. In: 1997 I.E. international conference on acoustics, speech, and signal processing (ICASSP'97), vol. 2, IEEE Computer Society, Munich

James DA, Young SJ (1994) A fast lattice-based approach to vocabulary independent wordspotting. In: International conference on acoustics, speech, and signal processing (ICASSP'94), IEEE CS, Adelaide

Jurafsky MJH (2000) Minimum edit distance. Speech and language processing: an introduction to natural language processing, computational linguistics, and speech recognition. Prentice Hall, Upper Saddle River

Kai T, Suzuki M et al (2012) Combination of SPLICE and feature normalization for noise robust speech recognition. In: International workshop on nonlinear circuits, communications and signal processing (NCSP'12), Honolulu

Kamm TM, Meyer GGL (2002) Selective sampling of training data for speech recognition. In: Proceedings of the second international conference on human language technology research, Morgan Kaufmann Publishers Inc, San Francisco

Kuri-Morales A, Rodríguez-Erazo F (2009) A search space reduction methodology for data mining in large databases. Eng Appl Artif Intel 22(1):57–65

Mammone RJ, Zhang X et al (1996) Robust speaker recognition: a feature-based approach. IEEE Signal Process Mag 13:58

Mamou J, Ramabhadran B (2008) Phonetic query expansion for spoken document retrieval. In: Interspeech'08, Brisbane

Mamou J, Ramabhadran B et al (2007) Vocabulary independent spoken term detection. In: 30th annual international ACM SIGIR conference on research and development in information retrieval, ACM, Amsterdam

Mamou J, Mass Y et al (2008) Combination of multiple speech transcription methods for vocabulary independent search. In: Workshop on search in spontaneous conversational speech (SIGIRSSCS'08), Singapore

Matrouf D, Gauvain J-L (1997) Model compensation for noises in training and test data. In: 1997 I. E. international conference on acoustics, speech, and signal processing (ICASSP'97), IEEE Computer Society, Munich

Mishne G, Carmel D et al (2005) Automatic analysis of call-center conversations. In: The 14th ACM international conference on information and knowledge management, vol 2001, Stuttgart, pp 423-429

Moallem P, Faez K (2001) Search space reduction in the edge based stereo correspondence. In: 6th international fall workshop on vision, modeling, and visualization, Bremen

Motlicek P, Valente F et al (2012) Improving acoustic based keyword spotting using LVCSR lattices. In: International conference on acoustic speech and signal processing, Japan

Murveit H, Butzberger J et al (1993) Large-vocabulary dictation using SRI's DECIPHER speech recognition system: progressive search techniques. In: IEEE international conference on acoustics, speech, and signal processing (ICASSP'93), IEEE, Minneapolis

Ney H, Mergel D et al (1992) Data driven search organization for continuous speech recognition. IEEE Trans Signal Process 40(2):272–281

Padmanabhan M, Ramaswamy G et al (1998) Voicemail Corpus I. Linguistic Data Consortium (LDC), Philadelphia

Padmanabhan M, Kingsbury B et al (2002) Voicemail corpus, Part II. Linguistic Data Consortium (LDC), Philadelphia

Parada C, Sethy A et al (2010) Balancing false alarms and hits in spoken term detection. In: IEEE international conference on acoustics, speech and signal processing (ICASSP'10), IEEE, Dallas

Park Y, Patwardhan S et al (2008) An empirical analysis of word error rate and keyword error rate. In: The international conference on spoken language processing (ICSLP), Brisbane

Pucher M, Türk A et al (2007) Phonetic distance measures for speech recognition vocabulary and grammar optimization. In: 3rd congress of the Alps Adria Acoustics Association, Graz

Ramasubramanian V, Paliwal KK (1992) An efficient approximation-elimination algorithm for fast nearest-neighbor search based on a spherical distance coordinate formulation. Pattern Recognit Lett 13(7):471–480

Richardson F, Ostendorf M et al (1995) Lattice-based search strategies for large vocabulary speech recognition. In: International conference on acoustics, speech, and signal processing (ICASSP'95), vol 1. IEEE, Detroit, pp 576–579

Sankar A, Lee CH (1996) A maximum-likelihood approach to stochastic matching for robust speech recognition. IEEE Trans Speech Audio Process 4(3):190–202

Saon G, Chien J-T (2012) Large vocabulary continuous speech recognition recognition systems. IEEE Signal Process Mag (IEEE Signal Processing Society) 29:18–33

Saraclar M, Sproat R (2004) Lattice-based search for spoken utterance retrieval. In: HLT-NAACL (2004), Boston

Schneider D (2011) Holistic vocabulary independent spoken term detection. Ph.D. dissertation. Rheinischen Friedrich-Wilhelms-Universitaat Bonn, Bonn

Schwartz R, Austin S (1991) A comparison of several approximate algorithms for finding multiple (N-best) sentence hypotheses. IEEE

Schwartz R, Chow YL (1990) The N-best algorithms: an efficient and exact procedure for finding the N most likely sentence hypotheses. In: 1990 international conference on acoustics, speech, and signal processing, 1990. ICASSP-90. IEEE, Albuquerque, pp 81–84

Schwartz R, Austin S et al (1992) New uses for the N-best sentence hypotheses within the BYBLOS speech recognition system. In: 1992 international conference on acoustics, speech, and signal processing, 1992. ICASSP-92,vol 1. IEEE, San Francisco, pp 1–4

Seide F, Yu P et al (2004) Vocabulary-independent search in spontaneous speech. In: IEEE international conference on acoustics, speech, and signal processing (ICASSP'04), Montreal

Šmídl, L, Psutka J (2006) Comparison of keyword spotting methods for searching in speech. In: Interspeech 2006, ISCA, Bonn

Srinivas M, Patnaik L (1991) Learning neural network weights using genetic algorithms-improving performance by search-space reduction. IEEE

Szöke I, Schwarz P et al (2005) Comparison of keyword spotting approaches for informal continuous speech. In: Eurospeech'05, Lisbon

Szöke I, Fapšo M et al (2008) Spoken term detection system based on combination of LVCSR and phonetic search. In: The 4th international conference on machine learning for multimodal interaction, Springer, Berlin

Tetariy E, Aharonson V et al (2010) Phonetic search using an anchor-based algorithm. In: Proceedings of IEEE 26th convention of electrical and electronics engineering in Israel, Eilat

Tetariy E, Gishri M, Har-Lev B, Aharonson V, Moyal A (2012) An efficient lattice-based phonetic search method for accelerating keyword spotting in large speech databases. Int J Speech Technol (2012):1–9

Thambiratnam K (2005) Acoustic keyword spotting in speech with applications to data mining. PhD, Speech and Audio Research Laboratory of the SAIVT Program – Center for Built Environment and Engineering Research. Queensland University of Technology, Brisbane, p 248

Thambiratnam K, Sridharan S (2005) Dynamic match phone-lattice searches for very fast and accurate unrestricted vocabulary keyword spotting. In: IEEE international conference on acoustics, speech, and signal processing (ICASSP'05), Philadelphia

Thambiratnam K, Sridharan S (2007) Rapid yet accurate speech indexing using dynamic match lattice spotting. IEEE Trans Audio Speech Lang Process 15(1):346–357

Tsao Y, Li J et al (2009) Ensemble speaker and speaking environment modeling approach with advanced online estimation process. In: IEEE international conference on acoustics, speech and signal processing (ICASSP'09), IEEE Computer Society, Taipei

Vergyri D, Shafran I et al (2007) The SRI/OGI 2006 spoken term detection system. In: 8th annual conference of the international speech communication association (INTERSPEECH 2007), ISCA, Antwerp

Viikki O, Laurila K (1998) Cepstral domain segmental feature vector normalization for noise robust speech recognition. Speech Commun 25(1):133–147

Wallace R, Vogt R et al (2007) A phonetic search approach to the to the 2006 NIST spoken term detection evaluation. In: 8th annual conference of the international speech communication association (INTERSPEECH 2007), ISCA, Antwerp

Wang, D, Tejedor J et al (2008) A comparison of phone and grapheme-based spoken term detection. In: IEEE international conference on acoustics, speech and signal processing (ICASSP'08), Las Vegas

Wilpon JG, Rabiner LR et al (1990) Automatic recognition of keywords in unconstrained speech using hidden Markov models. IEEE Trans Acoust Speech Signal Process 38(11):1870–1878

Witbrock MJ, Hauptmann AG (1997) Using words and phonetic strings for efficient information retrieval from imperfectly transcribed spoken documents. In: The second ACM international conference on digital libraries. ACM, Philadelphia, pp 3–35

Young SJ (1993) The HTK hidden Markov model toolkit: design and philosophy. Technical Report TR 153, Department of Engineering, Cambridge University, Cambridge

Young S (1996) A review of large-vocabulary continuous speech recognition. Signal Process Mag IEEE 13(5):45

Young SJ, Russell N et al (1989) Token passing: a simple conceptual model for connected speech recognition systems. Engineering Department, Cambridge University, pp 1–23

Young SR, Hauptmann AG et al (1989b) High level knowledge sources in usable speech recognition systems. Commun ACM 32(2):183–194

Yu P, Seide F (2004) A hybrid word/phoneme-based approach for improved vocabulary-independent search in spontaneous speech. In: INTERSPEECH 2004, Korea

Žgank A, Horvat B et al (2005) Data-driven generation of phonetic broad classes, based on phoneme confusion matrix similarity. Speech Commun 47(3):379–393